DIGITAL INFORMATION LITERACY IN AGRICULTURAL UNIVERSITIES

The Author

Dr. Yogita Sharma has a brilliant academic career. She has a Bachelor and Masters Degree in Library and Information Science from Panjab University, Chandigarh with gold medal. She has done her PhD from Indira Gandhi National Open University, New Delhi. She is in this profession for the last nineteen years. Presently, she is working as Deputy Librarian and heading the Digital Library along with the Technical Division at Punjab Agricultural University Library, Ludhiana. Dr. Yogita Sharma has published various research papers in professional journals and presented a number of papers in various conferences and seminars. She also teaches Library and its services course to the post-graduate students in PAU. She is associated with various professional associations. Her areas of interest are digital information literacy, e-resources, information science, library management and digitization.

DIGITAL INFORMATION LITERACY IN AGRICULTURAL UNIVERSITIES

The Author

Dr. Yogita Sharma has a brilliant academic career. She has a Bachelor and Masters Degree in Library and Information Science from Panjab University, Chandigarh with gold medal. She has done her PhD from Indira Gandhi National Open University, New Delhi. She is in this profession for the last nineteen years. Presently, she is working as Deputy Librarian and heading the Digital Library along with the Technical Division at Punjab Agricultural University Library, Ludhiana. Dr. Yogita Sharma has published various research papers in professional journals and presented a number of papers in various conferences and seminars. She also teaches Library and its services course to the post-graduate students in PAU. She is associated with various professional associations. Her areas of interest are digital information literacy, e-resources, information science, library management and digitization.

DIGITAL INFORMATION LITERACY IN AGRICULTURAL UNIVERSITIES

YOGITA SHARMA

Deputy Librarian
Mohinder Singh Randhawa Library
Punjab Agricultural University
Ludhiana (India)

SCIENTIFIC
PUBLISHERS

Published by:

Scientific Publishers (India)
5 A, New Pali Road, P.O. Box 91
Jodhpur - 342 001 (India)

Branch Office
Scientific Publishers (India)
4806/24, Ansari Road, Daryaganj
New Delhi - 110 002 (India)

E-mail: info@scientificpub.com
Website: www.scientificpub.com

ISBN: 978-93-83692-50-7

Visit the Scientific Publishers (India) website at
http://www.scientificpub.com

Printed in India

Digitally Printed at Replika Press Pvt. Ltd.

PREFACE

The libraries are basically service institutions with users and collection constituting the major ingredients. Undoubtedly, users form the backbone and reason for the existence of the libraries. The main focus of the libraries is to serve the users' upto their satisfaction. Consequently, the user studies are an integral component of research in library and Information Science in India, as such studies enable the Libraries to improve further and serve the users better. It is a known fact that the libraries have changed gradually to keep pace with technological advancements in terms of automation, digitization and digital resources of information, which has become a dominating part of the collection and services of the Libraries nowadays. Accordingly, it is of dire need to study the adaptability of users' towards digital resources of information. The same paved my way into initiating and successfully completing a doctoral research work on digital information literacy of post-graduate students of the agricultural universities of Punjab, Haryana and Himachal Pradesh. This book aimed to share my doctoral research work with wider professional community.

The present study has been conducted using relevant investigation methods and data collection tools and techniques. Descriptive research design with survey method was used with a fact-finding approach. The research methodology followed in this study involved collection and interpretation of data to find causal connections and relations for the existing status of DIL skills on sample of students. Further, the questionnaire method has been adopted as a data collection tool.

A systematic attempt is made to determine various aspects like extent of use of digital resources of information (DRI), proficiency in searching and evaluating DRI, understanding of ethical use of DRI, adoption of DIL in social lives, training needs with regard to DIL and role of the library and information science professionals in the information gathering process from DRI. Questions were framed about the digital information requirements of the students, how effectively and efficiently students locate, evaluate and use information from DRI. In order to enable cross-checking of data for its validity, reliability and consistency, the data was analyzed for various statistical parameters-frequency distribution, percentage, weighted index, Karl Pearson's Correlation Coefficient and Pearson's Chi Squared Test. Some related past studies were also consulted and an extensive comparison of the results was made. However, review of literature has been merged with introduction chapter in concise form. The results and findings have been presented elaborately and further areas of research have also been identified.

This book intends to serve as an important tool for students, teachers, researchers and practitioners in the field of Library and Information Science. This book will be highly useful to the researchers in terms of methodology and findings to conduct similar studies on other groups of users in the country and comparing similar studies done abroad.

I will always remain indebted to my mentor and supervisor of my research work Prof. Neena Talwar Kanungo, Faculty of Library and Information Science, School of Social Sciences, Indira Gandhi National Open University, New Delhi for her valuable guidance and never ending support. I am grateful to the students of Chaudhary Charan Singh Haryana Agricultural University, Hisar; Chaudhary Sarwan Kumar Himachal Pradesh Krishi Vishvavidyalya, Palampur and Punjab Agricultural University, Ludhiana for extending their whole hearted support by filling the questionnaires. I am obliged to the authorities and colleagues of these universities who have helped me in this research work. My thanks are also due to Scientific Publishers (India), Jodhpur for publishing the book.

YOGITA SHARMA

Ludhiana

LIST OF ABBREVIATIONS

Sr. no.	Term	Abbreviation
1	American Library Association	ALA
2	Association of College and Research Libraries	ACRL
3	Automatic Teller Machine	ATM
4	Chaudhary Charan Singh Haryana Agricultural University, Hisar	HAU
5	Chaudhary Sarwan Kumar Himachal Pradesh Krishi Vishvavidyalaya, Palampur	HPKV
6	Chi-squared test/ Pearson's chi-squared test/ Chi-square/ Pearson's chi-square	χ^2
7	Commonwealth Agricultural Bureaux	CAB
8	Consortium for e-Resources in Agriculture	CeRA
9	Correlation/ Correlation coefficient/ Karl Pearson's correlation coefficient	r
10	Degree of freedom	df
11	Digital Information Competency Level	DICL
12	Digital Information Competency Level Score	DICLS
13	Digital Information Literacy	DIL
14	Digital Resources of Information	DRI
15	Figure	Fig.
16	Indian Council of Agricultural Research	ICAR
17	Information and Communication Technology	ICT
18	Information Literacy	IL
19	Library and Information Science	LIS

20	Number of Respondents	R
21	Post-graduate	PG
22	Punjab Agricultural University, Ludhiana	PAU
23	Under-graduate	UG
24	Uniform Resource Locator	URL
25	Weighted Index	W.I.

LIST OF TABLES

LIST OF FIGURES

CONTENTS

INTRODUCTION

1

1.1 BACKGROUND

The world is flooded with information these days and its extent is going to increase even more rapidly in near future. It can rightly be asserted that fastest growing magnitude nowadays is the volume of information. It has been expanding faster than anything else being created. The information is being produced in print as well as in electronic form for example, e-books, e-journals, databases, bibliographic databases, reference works published online and other web-based resources. More new information has been produced within the last three decades than in the last five millennia. More than 1,000 books are published daily around the world. The Internet is spreading information much faster than printed works. In fact, the Internet, so far is one of the fastest growing communication media in world history. It has taken World Wide Web just 4 years to achieve an audience of 50 million users, compared to 38 years for radio, 16 years for personal computers, and 13 years for television (McCarthy). In 2007, the amount of digital information created, captured, and replicated was 2.25 x 10^{21} bits. In computer parlance, that is 281 exabytes or 281 billion gigabytes. This is about 5 million times the information in all the books ever written (Gantz). The problem of lack of information for centuries transformed into lack of access to information some decades ago but the information revolution has turned the tables and the problem now being faced is how to find relevant and correct information. This forms the basis of the need for information literacy (IL) which has been defined by Association of College and Research Libraries (ACRL) as 'a set of abilities requiring individuals to recognize when information is needed and have the ability to locate, evaluate and use effectively the needed information'. Gradually, with the advent of digital resources of information (DRI) the concept of digital information literacy (DIL) came into existence.

1.2 Digital Information Literacy- Historical Perspective

The word information is derived from Latin word *information,* which means **concept** or **idea**. The word literate in Latin means *learned or lettered.* The conventional definition of the term 'literacy' was ability to read and write at a basic level and sign a document. Over the time this definition changed and the United States (US) Workforce Investment Act, 1998 defined literacy as 'an individual's ability to read, write, speak in English, compute and solve problems at levels of proficiency necessary to function on the job, in the family of the individual and in society' (Welsh and Wright 1-11). The term **information literacy** was coined in 1974 by Paul Zurkowski. Its origins were not specifically located in higher education. Zurkowski used the phrase to describe the techniques and skills known by the information literate "for utilizing the wide range of information tools as well as primary sources in molding information solutions to their problems." Rockman views IL as "truly a new instructional pedagogy and a change agent for learning" (47-91). Bewick and Corrall reported that professional interest in IL is several decades old, though earlier different terms were used for it, such as bibliographic or library instruction and user education (97-110).

The concept of IL was conceptualized long back in European countries during 1970s and recognized as critical literacy for the 21st century. The developed countries realized its importance in 1980s and by 1990s they took concrete steps in that direction. In United States of America (USA), the National Forum of Information Literacy was established in 1989 and the Institute of Information Literacy in 1998. The US Department of Education included IL in National Education Technology Plan as one of the five goals in December 2000. There are many IL organizations in USA which impart training in IL and also develop standards for the same. Some of these are:

1. ACRL's Instruction Section
2. American Library Association (ALA)'s Library Instruction Round Table
3. Library Orientation Exchange
4. California Clearinghouse on Library Instruction
5. International Federation of Library Associations and Institutions (IFLA) Information Literacy Section (Prasher and Sharma 23-47).

The academicians and practitioners around the world have articulated different frameworks and models to support IL development with notable examples including Bruce's 'seven faces' model, the Society of College National and University Libraries' (SCONUL) 'seven pillars', and the ACRL's competency standards. The Information Literacy Competency Standards for Higher Education given by ACRL form the basis of present study and are given as below:

1. The information literate student determines the nature and extent of the information needed.
2. The information literate student accesses needed information effectively and efficiently.
3. The information literate student evaluates information and its sources critically and incorporates selected information into his or her knowledge base and value system.
4. The information literate student, individually or as a member of a group, uses information effectively to accomplish a specific purpose.
5. The information literate student understands many of the economic, legal, and social issues surrounding the use of information and accesses and uses information ethically and legally (Association of College and Research Libraries).

Information literacy competency standards states that IL is common to all disciplines and forms the basis for lifelong learning (Association of College and Research Libraries 2). These standards also advocate that IL extends learning beyond formal education where information literate individuals can apply these skills in their jobs and social lives with information management as hub of all activities (Hepworth 21-34; Johnston and Webber 335-52; Lindstrom and Shonrock 18-23; Stubbings and Brine).

The urge to have speedy access to the required information has brought out a revolution in preferences of information seekers. The information via DRI is easily and abundantly available within few seconds. In other words, a simple touch of button expeditiously provides access to bounty of information. The skills of an individual to locate, evaluate, organize and use information effectively are tested to the limits while churning the ocean of DRI. Krishnamurthy refers to sources in electronic or digital format as DRI or e-resources (245-54). DRI have exploded in popularity and usage because they enable innovation in teaching, improves timeliness in research, and creation of new ideas.

These resources provide up-to-date information which is sometimes not available elsewhere. Gradually, with the emergence of DRI the focus of all such organizations was on development of competency skills enabling effective use of such resources. The initial concept of literacy, which as we have seen goes considerably beyond a simple ability to read and write, has been complemented in recent decades as 'skill-based literacy's' that includes library literacy, media literacy and computer literacy. These concepts have been developed to deal with information of increasing complexity, and developing technologies. This skill-based literacy's have emerged to meet the needs of a more complex information environment with new technologies and a wider variety of media and services. Centered on a core of skills, all of them moved beyond this, showing that, as with literacy itself, that these seemingly simple forms of literacy require a wide spectrum of skills, knowledge, understanding and attitudes. This leads us on to a consideration of a form of literacy appearing from the outset to be based on rather newer premises.

As the world progressed from the information age to digital age, new definitions of IL started evolving that include how to access, evaluate and use information in digital formats (Welsh and Wright 1-11). Since Information and Communication Technology (ICT) has become widely adopted, IL has become a crucial component of learning and teaching in e-environments. There has been a trend towards adapting IL to dynamic e-environments that are environments in which the forms, content, and ways of searching, accessing, evaluating and using information sources and services are evolving (Andretta, *From Prescribed Reading*; Bridgland and Whitehead 54-9; Doherty, Hansen, and Kaya 1-12; Hadengue 36-46; Joint, *Information Literacy Evaluation: Moving Towards Virtual Learning Environments*; Kennedy; Ray and Day; Spitzer, Eisenberg, and Lowe; Todd). Several authors identify IL as inadequate for complex and evolving context of e-environments (Martin, *Essential e-literacy* 22-4; Savolainen 211-26; Williams, *Opinion against Information Literacy* 20-2). According to Kokic the traditional notion of IL is inadequate to address the learning needs of students in the 21st century thereby leading to evolution of the concept of digital information literacy (45-55).

The prevalence of information in digital form has led to the emergence of DIL which is strictly focused to effectively locating and using the DRI. Lanham defined DIL as the ability to understand and use information in multiple formats from a wide variety of networked computer sources and it involves the skill of deciphering multimedia images, sounds and text (160-61). According to Sieberhagen DIL in simpler terms is 'a set of digital abilities requiring individuals to recognize when digital information is needed and have the ability to locate, evaluate, organize and effectively use the needed digital information in an ethical and legal way.' Thus, DIL is a major component of IL; however, the pre-requisite for the former is one's proficiency in computer and Internet. DIL also enlightens a person regarding issues like copyright, plagiarism and intellectual property rights, etc. in an electronic environment.

Thus, the foundation for DIL is the broader library-instruction movement developed over more than 100 years of experimentation and implementation. Most of the libraries and information systems introduce user orientation programs that educate users on the salient features of information sources, search techniques, search strategies, scholarly communication and other aspects. With technology taking a lead in dissemination of information, changes were forced on library instruction programs leading to introduction of IL courses with special emphasis on DRI that had a wholesome approach to improve competencies.

Digital information literacy, along with similar other literacies such as web literacy, gained currency in the mid-1990s (Bawden, *Information and Digital Literacies* 218-59; Godwin and Parker). Initially, the libraries emphasized IL as an overarching concept and web literacy was considered as a subset of IL. Besides the library sector which had invested significantly over intellectual and strategic resources in IL, other literacy constituencies, notably the academic domains of education and cultural studies, gave DIL a distinctive theoretical shape (Owusu-Ansah 219-30). The most remarkable contrast with IL was a shift from the consumption to the production of information and expressive culture i.e. from 'read only' to 'read-write' modes (Hartley 17). This contributory or participatory aspect was theorized before the development of Web 2.0 or social media tools. However, the low 'entry costs' of Web 2.0 brought the concept of participatory culture and the skills of DIL that underpin it, to wide attention (McShane 383-93).

Web 2.0 is commonly associated with web applications that facilitate interactive information sharing, interoperability, user-centered design, and collaboration on the World Wide Web. It thus allows users to interact with one another or edit or add to website content. Web 2.0 and the growth in other participatory technologies made a remarkable impact on the information environment. Changes in the information environment also require a corresponding change in the way IL is conceptualized and taught. Students need to develop new literacies/skills to be successful in the 2.0 classroom and in the 2.0 world. Nowadays, the skills required for an individual to successfully find, evaluate and use information are different from what they were a decade ago when the ACRL Information Literacy Competency Standards for Higher Education were developed and students need to be prepared for a world that requires continuous informal learning. Therefore, IL must change to match the current educational and technological environment (Farkas 82-94).

Digital information literacy is progressively more important in the existing environment of rapid technological change and proliferating information resources. Because of the growing complexity of this environment, individuals have diverse and abundant information at their disposal. Information is available through libraries, community resources, special interest organizations, media and internet. Information comes to individuals in unfiltered formats, raising doubts about its authenticity, validity and reliability. Additionally, information is available through multiple media, including graphical, aural and textual and these creates new challenge for individuals in evaluating and understanding it (American Library Association).

A person needs to possess specific skills in order to use DRI efficiently. Rowlands and Bawden support the viewpoint that learning in digital library environment encompasses two distinct components. Firstly, learning how to access, evaluate and use information resources and secondly, learning how to master and build on the ideas embodied within resources (192-202). Wilson has defined the current state as a state of data smog that makes traditional literacy alone inadequate for survival in the digital world. Therefore, while DIL shares some of the critical attributes of literacy in the traditional sense, it requires additional competencies. Computer literacy, media literacy and network literacy are among the prerequisite skills that are required for users to be digitally information literate (189-95).

Digital information literacy needs to be developed throughout life by information users, librarians and information workers. The need of DIL is more in case of higher education because higher education institutions provide students with both problem-solving and decision-making skills that enable them to compete in their workplaces. In order to achieve this, it is of dire need to prepare students to be information literate individuals with respect to digital environment and possess the ability to locate process and use information from DRI effectively and become excellent producers and consumers of information (Prasher and Sharma 23-47).

Maharana and Mishra conducted a survey upon the faculty of PG departments at Sambalpur University with an objective to study the awareness about DIL amongst the faculty and further set priorities for promotion of DIL. The study revealed an educational imbalance between the ever-developing technologies and information available to users. Furthermore, a dire need to educate people regarding use of DRI was felt.

Nowadays, every literate person must achieve a level of DIL to accompany traditional literacy efforts. The demand for such contemporary literacy skills to match with traditional literacy presents an opportunity for university based instructional designers to share expertise and to collaborate with their local communities to this end (Tryon 52-58).

DIL is not only confined to using digital information efficiently and effectively but also to use it in an ethical and legal way is most imperative. DRI specifically Web 2.0 technologies have the communicative potential of the Internet that has become increasingly accessible and enables users with the power to publish multimodal texts. These technological tools are misused, often marginally by people to become publishers of information (Gainer 14-17).

Joint further emphasizes that in this knowledge economy, librarians and informational professionals shoulder an exceptional responsibility and that is of framing a strategy, which includes understanding and resolving legal disputes and conflicts unique to the unethical use of information from digital resources. This mark may be best achieved by including intellectual property issues in syllabus of DIL. (*Teaching Intellectual Property Rights as Part of the Information Literacy Syllabus* 330-36).

Lampert stresses upon developing a discipline-based approach to the issue of plagiarism in IL instruction programmes. In this cut and paste age, author intends to examine how instances of academic dishonesty can be prevented in order to prepare our students to use information ethically in their professional lives beyond the college experience. Librarians and discipline faculty should teach the students how to acknowledge and cite the information used (347-55).

It has been realized that DIL skills are mandatory for an individual in order to survive in an ever exploding world of information. To achieve the objective of digitally information literate society various ways are being explored by the experts. The efforts of achieving this goal by mere experience of an individual and education imparted by the professionals were found to be insufficient. It was felt that imparting DIL education as a non-credit course, as prevalent in various institutions, is an ambiguity in success. Thus, educational institutions worldwide are being encouraged to incorporate DIL programmes in their curriculum as a compulsory course. Dadzie opines that in order to make sure that all students acquire competencies in knowing how to learn, to formulate questions, to access potential sources of information, to evaluate what is found for accuracy, to organize information and finally to use information; it is imperative for the universities to incorporate the concept of DIL in the curricula. Hence, it will be certain that individuals are adequately equipped by the time they complete university education (266-77). Integration of DIL into course curriculum can inculcate research strategies and promote effective use of print and digital resources. It is expected from the students, the education community and supervisors of tomorrow, to be superior consumers of information, contemplating the efficiency, effectiveness, and ethics of information, whether in print or in digital form. It has been observed that students who are able to use computers are not necessarily able to use the digital information to best effect. Leigh and Gibbon have recommended that librarians and faculty should redesign curriculum and incorporate mandatory DIL part into it. A proper and compulsory education provided to students will develop their DIL skills. (509-30).

This view has been supported by many research communities. Chakraborty, Das and Sen conducted a study on biomedical practitioners of the West Bengal region to find out the extent of use of open access biomedical literature; and their awareness about the biomedical databases, electronic journals, digital libraries, digital archives and Web portals of biomedical organizations. The study revealed under-utilization of open access biomedical information resources. It was recommended that the awareness level of the respondents may be increased through IL programmes on use of DRI. It was also recommended that the curricula of under-graduate (UG) and post-graduate (PG) programmes in biomedical sciences should include DIL skills that ensure lifelong learning. This will enable the practitioners to become competent to use new information and are able to apply the new knowledge in their professional life for the better benefits of the society.

Nelson, Courier, and Joseph described the results of a survey of teaching faculty at a medium-sized university in the southeastern US to determine core curriculum items that should be taught to ensure that graduates have the capabilities and skills to fully participate in the digital era. Application skills continued to be viewed as relevant. However, the findings show the need to focus on DIL skills as compared to the current one-credit-hour software applications course on computer application literacy. Additional topics may also be included in the curriculum common to all students such as ethics, security and privacy and how to validate the relevance and usefulness of data. The study also highlights the need for discipline-specific topics to be embedded in subject-knowledge courses (95-109).

Educating the students regarding use of information available in digital resources, simply by including it in the curriculum is not enough to inculcate DIL skills amongst them. There are other factors associated with developing these skills amongst the students. Librarians and faculty are two such vital factors. The library staff as well as the teachers can play a remarkable role in making the students digitally information literate. Librarians may teach the effective use of information both from various DRI available in libraries as well as not available in libraries. This effort will be more rewarding if faculty collaborates with librarians. Faculty can add subject specific ingredients to this training. Thus, students would definitely develop DIL skills which are useful not only in their academic lives but in their social lives also.

Librarians have always been acknowledged for their role in educating people by promoting the use of libraries even when availability of information was restricted only to the books available in libraries. Gradually, media changed its shape to digital and information too became widely available from sources other than libraries like the Internet. The transformation in media or sources of information left no effect on the role of librarians. Apart from assisting the individuals in accessing the required information they cultured them to become self–reliant in using it. It has been realized that DIL is a life-long learning process and imbibing this competency along with other associated technologies, like use of computers, amongst the people is a major challenge ahead for the librarians.

Today's students are overusing Google and Wikipedia instead of relying on the scholarly and authoritative research work that can be found through the virtual library databases of which they are authenticated users. The librarians need to understand the new tools, the new databases, the new searching capabilities, the new bibliographic management software, the new ways of assessing and teaching DIL skills; and the broad and differentiating backgrounds of their students' DIL skills. At another level they have to be more proactive than they have in the past; they need to speak powerfully, persuasively and credibly about student learning outcomes, along with being able of working with colleagues across campus to help faculty make modern DIL instruction part of their teaching (Lorenzo, *Catalysts for Change: Information Fluency, Web 2.0, Library 2.0, and the New Education Culture*).

McCarthy opines that availability of information in DRI has made DIL skills a necessity for all, so that the abundantly available information can be used efficiently. He stressed that the skills such as problem solving skills, thinking skills, communication skills, computer skills, etc. acquired by use and practice, will enable the person to compete with information-based society. In order to make students digital information literate the role of the teachers needs to be transformed to that of a facilitator or a coach.

IL and DIL are central topics for the information science. They are associated with issues as varied as information overload, lifelong learning, knowledge management and the growth of the information society. Naturally, they have been much discussed in the literature, but not, perhaps, as much as their importance deserves; in particular, they have not impinged much on the practitioner. To deal with the complexities of the current information environment, a complex and broad form of literacy is required. It must subsume all the skill-based literacy's, but cannot be restricted to them, nor to any particular technology or set of technologies. Understanding, meaning and context must be central to it. It is not much importance whether this is called IL, DIL, or simply literacy for an information age. Its importance can be actively promoted as a central core of principles and practice of the information sciences. It is thus concluded that since the need of DIL has arisen due to rapid increase of information in digital form and its wide dispersal facilitated by ICT. The informational needs of users have also increased manifold keeping pace with information explosion as well as ongoing research on complex and inter disciplinary areas. Presently, IT has emerged as an indispensable component of higher education. For optimum exploitation of information exploding in digital forms DIL needs to be enhanced, perhaps by integrating with academic activities. This will in turn lead to improvement in quality of education and research.

1.3 Definitions of Digital Information Literacy

Digital information literacy has been widely considered as an essential requirement for life in a digital age. According to Martin, DIL is the "awareness, attitude and ability of individuals to appropriately use digital tools and facilities to identify, access, manage, integrate, evaluate, analyze and synthesize DRI, construct new knowledge, create media expressions, and communicate with others, in the context of specific life situations, in order to enable constructive social action; and to reflect upon this process" (*Literacies for the Digital Age* 3-25).

Romani and Juan has defined DIL as integration of instrumental skills (information management) and strategic skills (critical thinking) to create, adapt and share information/ knowledge in multiple formats. Martin and Grudziecki identified three levels or stages for DIL development: the lower stage of digital competence (skills, concepts, approaches, attitudes, etc.), the central and crucial level of digital usage (application of digital competence within specific professional or domain contexts), and the ultimate stage of digital transformation where digital usages are developed to "enable innovation and creativity, and stimulate significant change within the professional or knowledge domain" (249-67).

In nutshell, many definitions of DIL appear to be built on three principles: the skills and knowledge to use a variety of digital media ...; the ability to critically understand digital media content ....; and the knowledge and capacity to create new information using digital technology and content... (Media Awareness Network). As such, the concept of DIL is much broader than computer literacy/ICT literacy, and instead represents an umbrella framework for integrating other inter-related sub-disciplines/literacies and skill-sets such as technology literacy, IL, media literacy and visual literacy (Covello; Bawden, *Origins and Concepts of Digital Literacy*).

1.4 Digital Information Literacy Research in India

The importance of IL was recognized in developed countries in 1990s and since then they have been working hard in its implementation and development. Most of the work regarding DIL has been done in USA, Australia, New Zealand and some European countries. DIL is still an ambiguity for the people of developing countries specially Middle East and North Africa. Fahmy and Rifaat reported that lack of indigenous Arabic content materials on the Internet is related to the challenges of increasing the awareness of the people of Middle East and North Africa with reference to DIL (111-23). In India, very little has been done in this direction and that too in form of seminars, conferences and workshops to use specific databases or e-resources. India needs to take concrete steps to conceptualize it. Consequently, it will optimize the efficient use of digital information and improve the quality of our education results an informative and learned India. Prasher and Sharma have suggested that:

1. India should develop a National Forum on IL.
2. Concrete policies should be made for implementation of IL.
3. IL should be implemented in phased but fast-track manner at all levels.
4. India should not adopt International IL Standards and should also develop its own standards to meet the local variations/demands.
5. IL should be incorporated in curricula at all levels of education.
6. India has realized the potential of distance education; it has to now develop an IL programme to specially meet the requirements of distance learners.
7. The training of library professionals and academicians as IL instructors should be taken up on war footing (23-47).

In light of the above, it is suggested that since today's world is flooded with information available in DRI so all the above suggested efforts may be focused on using information efficiently and effectively from DRI or in other words to the implementation and development of DIL skills.

Gedam and Agashe have discussed about IL programmes conducted by the government in India and observed that no significant steps have been taken by the government in this direction. It was reported that such programmes are simply literacy programmes and not IL programmes. However, Indian Institutes of Technology, Indian Institutes of Management and other Universities conducted IL programmes, but these are mostly conducted by the publishers/ aggregators and thus, are restricted to using their databases and e-resources. It was suggested that IL with specific focus on using DRI should be integrated with curricula at national level for all grades and also for lifelong learning (520-28).

Chauhan, Prem Chand and Murthy reported that Information and Library Network, Ahmedabad, has made various efforts like in-house training programmes, user awareness programmes, special training programmes and seminars to promote IL in higher education and research community and also for proper utilization of e-resources (45-52). In nutshell, DIL programmes in India can be considered at budding stage. Much of the Indian literature is based largely on subjective information and there is a dearth of research on DIL particularly in the case of agricultural education.

1.5 Need of the Study

As discussed above, DIL programmes in India are in the initial phase of development. Much of the Indian literature is based largely on anecdotal information and there is a paucity of research on assessing the DIL skills possessed by the individuals. Few studies have been conducted on the effectiveness of use of DRI and libraries, but serious efforts to assess overall process of information search from DRI and its use by the learners are lacking, particularly in the case of agricultural education system.

In most of the agricultural universities, role of library professionals is limited to the extent of user orientation programmes conducted for the first year students and teaching bibliography or library and its services part of course on Technical Writing for Master degree students and PhD scholars. These universities are lacking in creating awareness about the concept of DIL and evaluation of DIL skills possessed by the students. A collaborative approach by the library professionals and faculty to develop DIL competencies of the learners has not been taken up so far.

Keeping this in view, the present study was conducted to get a deep insight into the whole process of information gathering process from DRI by the learners. An attempt was also made to study the factors affecting the process in order to pave the way for developing the much needed collaborative approach. The present study "Digital Information Literacy in Agricultural Universities" focused to determine the extent of use of DRI, proficiency in searching and evaluating DRI, understanding of ethical use of DRI and use of DRI in social lives by Master degree students and PhD scholars of Chaudhary Charan Singh Haryana Agricultural University, Hisar (HAU); Chaudhary Sarwan Kumar Himachal Pradesh Krishi Vishvavidyalaya, Palampur (HPKV) and Punjab Agricultural University, Ludhiana (PAU). The study also explored the factors affecting the DIL skills of the students and ascertains their training needs to improve their DIL skills.

1.6 Scope of the Study

Research work in master and doctorate degrees in agricultural universities is an integral part of agricultural education. The researchers consult journals, abstracting sources, indexing sources, theses, reference books, etc. for their research work in addition to textbooks and class notes. Majority of such resources are available in electronic form nowadays and possession of DIL skills by the students is need of the hour. Keeping in view the informational requirements of Master degree students and PhD scholars, the present study was conducted on Master degree students and PhD scholars of HAU, HPKV and PAU.

1.7 Objectives of the Study

The objectives of the present study are:

1. To determine the extent of use of digital resources of information by the master degree students and PhD scholars.
2. To determine the proficiency of master degree students and PhD scholars in searching and evaluating the digital resources of information.
3. To ascertain the training needs of master degree students and PhD scholars with regard to digital resources of information.
4. To assess the master degree students and PhD scholars understanding of ethical issues regarding use of digital resources of information.
5. To assess the use of digital information literacy by the master degree students and PhD scholars in their social lives.

1.8 Hypotheses of the Study

The following hypotheses were formulated and tested:

1. There will be no association between digital information competency level and age of users.
2. There will be no association between digital information competency level and gender of users.
3. There will be no association between digital information competency level and university of users.
4. There will be no association between digital information competency level and class of users.

5. Libraries programmes are not sufficient to enhance the digital information literacy skills of their users.

1.9 Definitions and Scope of the Terms Used

The definitions and scope of various terms used in this study are given below in this section.

1.9.1 Master degree students

Master degree students refer to the master degree students of all the three universities. These include students registered in Master of Science (M.Sc.), Master of Technology (M.Tech), Master of Business Administration (MBA), Master of Veterinary Science (M.V.Sc.) and Master of Computer Applications (MCA) degrees.

1.9.2 PhD scholars

PhD scholars refer to the doctorate degree students of all the three universities. These include students registered in Doctorate of Philosophy (PhD) degree programme of all the disciplines in all the three universities.

1.9.3 Post-graduate students

The master degree students and PhD scholars collectively are termed as Post-graduate students in all the three subject universities.

1.9.4 Class

The term class used in the present study refers to the class of study or degree of the respondent which includes master and doctorate degree. The master degree includes M.Sc., M.Tech, MBA, M.V.Sc. and MCA. The doctorate degree refers to PhD.

1.9.5 Agricultural universities

The state agricultural universities in India are governed by Indian Council of Agricultural Research (ICAR) and are established to co-ordinate agricultural education, research and extension activities in their respective states primarily in the fields of agriculture, agricultural engineering, home science, veterinary science and allied areas in basic science and humanities. At present, there are 100 ICAR institutes and 62 state agricultural universities across the country. The agricultural universities from the states of Punjab, Haryana and Himachal Pradesh have been covered under this study.

1.9.6 Credit course

A credit course often used as credit hour also, is a unit that gives weightage to the time requirements of an academic course taken at any educational institution. In agricultural universities, students receive credit hours based on the number of contact hours per week in class, for one semester. In such courses the students are evaluated in terms of marks which are added to their overall grading of the semester.

1.9.7 Non-credit course

A non-credit course in agricultural universities system refers to an academic course in which weightage in terms of marks is not given to the time requirements. Students in such course are graded as satisfactory or unsatisfactory remarks only.

1.9.8 Digital resources of information

The term Digital Resources of Information (DRI) has been used synonymous with e-resources which refer to sources available in electronic or digital form like online databases, CDs, e-journals, e-books and Internet. The DRI in the present study refers to various online and CDROM databases subscribed by all the three university libraries along with other relevant open access resources available through Internet.

1.9.9 Digital information literacy

Digital information literacy has been defined by several thinkers as the abilities possessed by the individuals to identify, access, integrate, evaluate, analyze, organize, create and communicate information from DRI. The focus has been laid on ethical use of DRI and adoption of DIL in their social lives. In the present study DIL refers to proficiency of the respondents in accessing, evaluating and ethically using DRI. It also refers to the adoption of DIL skills in their social lives.

1.9.10 Scores

Digital information competency level score

Digital Information Competency Level Score (DICLS) refers to a numerical value assigned to competency of the respondents in evaluating and using information from DRI.

Social score

Social score refers to a numerical value assigned to the adoption of DIL in the day to day life by the respondents.

Ethical use score

Ethical use score refers to a numerical value assigned to the understanding of ethical use and tendency to give due acknowledgement to the DRI used by the respondents.

Access score

Access score refers to a numerical value assigned to the frequency of access and amount of information accessed from DRI.

1.10 Limitations of the Study

Though every effort has been made to make the study as comprehensive as possible, but it is subject to limitations. Some of these limitations are given below:

1. Under-graduate students were excluded from the study because their information needs are usually limited to class notes and textbooks.
2. Department wise analysis of the population was not done due to inter-disciplinary nature of research and course contents in agricultural universities.
3. Every care was taken to minimize the errors due to ambiguity and other usual shortcomings of tools and techniques used for the study, still the possibility of errors cannot be completely ruled out.
4. As far as possible the questionnaires were filled in the presence of the investigator so as to avoid consultation with other persons. However, some questionnaires were returned back next day and a few were received by post thus the possibility of consultation cannot be completely ruled out.
5. The study is limited to agricultural universities of Punjab, Haryana and Himachal Pradesh therefore comprehensive results could not be obtained. The generalizations are applicable to agricultural universities only.

6. Findings of the study are based on the expressed opinions of respondents. Although every effort was made to get accurate information from respondents, the possibility of their giving some biased information cannot be completely ruled out.

1.11 Organization of the Study

The present study has been organized logically and effectively into eight chapters so that it may be convenient to consult it quickly at specific areas of interest. A list of references has been given at the end of every chapter. A brief outline of the chapters is given below.

1. Chapter I: Introduction: This chapter discusses the term DIL, its development and definitions in detail. The relevant literature consulted for this study has also been discussed in this chapter. Furthermore, it includes the need, statement, scope, objectives, hypotheses and limitations of the study. Definitions and scope of various terms used have also been included here.

2. Chapter II: Research Methodology: Various data collection and analysis techniques used in the study are described in this chapter. The different stages of data collection using questionnaire method which comprises of designing, pre-testing and administration of questionnaire have been included in this chapter. The response rate and various techniques used in data analysis and its interpretation have also been discussed in detail. The style of compilation of bibliography has also been talked about.

3. Chapter III: User Characteristics: This chapter comprises of demographic information of respondents. The data collected has been analyzed and presented therein on the basis of four variables of study namely age, gender, name of the university and class. It also consists of data analyzed and interpreted with regard to purpose of using and consultation of DRI. The perception and responsiveness of the respondents about ethical use of DRI has also been discussed in this chapter.

4. Chapter IV: Accessing, Searching and Selecting Digital Resources of Information: This chapter covers analyses and discussion of data for various aspects like accessing DRI, searching information from DRI and selecting authentic DRI.

5. Chapter V: Role of Librarians and Teachers: This chapter discusses the data analyzed regarding the support received by the users from librarians and teachers in using DRI. It also interprets the users' perception about the competency of the library staff in terms of their training with regard to DRI. Moreover, this chapter focuses on the problems faced by the users in using DRI.

6. Chapter VI: Digital Information Literacy in Social Lives: This chapter analyses the data and interprets it for adoption of DIL skills by the respondents in their day to day social life. Furthermore, the suggestions given by the respondents in improving the use of DRI have also been analyzed and interpreted therein.

7. Chapter VII: Digital Information Literacy Skills: This chapter discusses the correlation amongst various DIL skills namely DICLS, social score, ethical use score and access score. The correlation was calculated using Karl Pearson's correlation coefficient (r) and is presented in this chapter. In addition to this, the association between various DIL skills and different variables of study calculated using Pearson's chi-squared test (χ^2) has also been discussed.

8. Chapter VIII: Findings, Conclusions, Suggestions and Recommendations for Further Areas of Research: The findings, conclusions and suggestions of the study are given in this chapter. It also includes recommendations on areas for future research.

9. Bibliography: A bibliography of all the related works referred during the study has been compiled in Modern Language Association (MLA) style and constitutes the last part of the book.

10. Annexure: The questionnaire designed to collect data from the respondents is given in this part of the thesis.

REFERENCES

American Library Association. *Progress Report on Information Literacy: An Update on the American Library Association Presidential Committee on Information Literacy: Final Report. March 1998*. Chicago: American Lib. Assn., 1998. Web. 15 Nov. 2009.

Andretta, Susie. "From Prescribed Reading to the Excitement or Burden of Choice: Information literacy: Foundation of E-learning." *Aslib Proceedings: New Information Perspectives* 57.2 (2005): 181-90. *Emerald*. Web. 17 Sept. 2008.

Association of College and Research Libraries. *Information Literacy Competency Standards for Higher Education*. Chicago: American Lib. Assn., 2000. Web. 18 Oct. 2008.

Bawden, David. "Information and Digital Literacies: A Review of Concept." *Journal of Documentation* 57.2 (2001): 218-59. Print.

Bewick, Laura, and Sheila Corrall. "Developing Librarians as Teachers: A study of their Pedagogical Knowledge." *Journal of Librarianship and Information Science* 42.2 (2010): 97-110. *Sage.* Web. 24 Sept. 2011.

Bridgland, Angela, and Martha Whitehead. "Information Literacy in the 'E' Environment: An Approach for Sustainability." *Journal of Academic Librarianship* 31.1 (2004): 54–59. *ScienceDirect.* Web. 9 Oct. 2011.

Bruce, Christine Susane. *The Seven Faces of Information Literacy.* Adelaide: Auslib Press, 1997.Web. 29 May 2010.

Chakraborty, Susmita, Anup Kumar Das, and B. K. Sen. International Conference on Information Management in Digital Libraries. Aug. 2-4, 2006: Use of Indian Digital Libraries by Biomedical Practitioners in West Bengal Region: A Study. Kharagpur: n.p., 2006. Openmed.nic.in. Web. 18 Oct. 2008.

Chauhan, S. K., Prem Chand, and T. A. V. Murthy. "Information Literacy for Indian Academicians: INFLIBNET Initiatives." *Journal of Library and Information Science* 31.1 (2006): 45-52. *Sage Journals Online.* Web. 28 May 2012.

Covello, Stephen. *A Review of Digital Literacy Assessment Instruments.* N.p.: Syracuse Univ, 2010. Web. 4 April 2011.

Dadzie, Perpetua S. "Information Literacy: Assessing the Readiness of Ghanaian Universities." *Information Development* 23.4 (2007): 266-77. *Sage Journals Online.* Web. 23 July 2009.

Doherty, John J., Mary Anne Hansen, and Kathryn K. Kaya. "Teaching Information Skills in the Information Age: The Need for Critical Thinking." *Library Philosophy and Practice* 1.2 (1999): 1–12. *E-prints in Lib. and Information Sciences.* Web. 29 July 2009.

Fahmy, Engy I., and Nermine M. Rifaat. "Middle East Information Literacy Awareness and Indigenous Arabic Content Challenges." *International Information and Library Review* 42 (2010): 111-23. *ScienceDirect.* Web. 17 June 2011.

Farkas, Meredith. "Participatory Technologies, Pedagogy 2.0 and Information Literacy." *Library Hi Tech* 30.1 (2012): 82-94. *Emerald.* Web. 10 Feb. 2013.

Gainer, Jesse. "Critical Thinking: Foundation for Digital Literacies and Democracy." *Journal of Adolescent and Adult Literacy* 56.1 (2012): 14–17. *Wiley.* Web. 11 Feb. 2013.

Gantz, John F., et al. The Diverse and Exploding Digital Universe: An Updated Forecast of Worldwide Information Growth by 2011 (IDC White Paper). N. p., 2008. Web. 18 Oct. 2008.

Gedam, P. B., and A. T. Agashe. International Conference on Academic Libraries. Oct. 5-8, 2009: Information Literacy Competencies and Programmes in India: Proceedings. Delhi: n.p., 2009. Print.

Hadengue, Veronique. "E-learning for Information Literacy: A Case Study." *Library Review* 54.1 (2005): 36-46. *Emerald.* Web. 4 Sept. 2009.

Hepworth, Mark. "Approaches to Providing Information Literacy Training in Higher Education: Challenges for Librarians." *New Review of Academic Librarianship* 6.1 (2000): 21–34. *Taylor & Francis Online.* Web. 24 Aug. 2009.

Johnston, B., and S. Webber. "Information Literacy in Higher Education: A Review and Case Study." *Studies in Higher Education* 28.3 (2003): 335–52. *ingentaconnect.* Web. 20 Sept. 2009.

Joint, Nicholas. "Information Literacy Evaluation: Moving Towards Virtual Learning Environments." *The Electronic Library* 21.4 (2003): 322-34. *Emerald.* Web. 20 Aug. 2009.

Joint, Nicholas. "Teaching Intellectual Property Rights as Part of the Information Literacy Syllabus." *Library Review* 55.6 (2006): 330-36. *Emerald.* Web. 9 Sept. 2009.

Kennedy, Cathleen. Measuring Information Literacy: The 'Tool Literacy' Variable. 1998. Google. Web. 29 July 2009.

Kokic, Ivana Batarelo. "Information Literacy for Future Teachers." *World Journal of Education* 2.1 (2012): 45-55. *Sciedu Press.* Web. 4 Feb. 2013.

Krishnamurthy, M. "Digital Libraries Gateway for Library and Information Sciences: A Study." *SRELS Journal of Information Management* 39.3 (2002): 245-54. Print.

Lampert, Lynn D. "Integrating Discipline Based Anti-plagiarism Instruction Into the Information Literacy Curriculum." *Reference Services Review* 32.4 (2004): 347-55. *Emerald.* Web. 31 Oct. 2009.

Lanham, Richard A. "Digital Literacy." *Scientific American* 272.3 (1995): 160-61. Print.

Leigh, Jennifer S. A., and Cynthia A. Gibbon. "Information Literacy and the Introductory Management Classroom." *Journal of Management Education* 32.4 (2008): 509-30. *Sage Journals Online.* Web. 22 Oct. 2009.

Lindstrom, J., and D. D. Shonrock. "Faculty–Librarian Collaboration to Achieve Integration of Information Literacy." *Reference and User Services Quarterly* 46.1 (2006): 18–23. Print.

Lorenzo, George. Catalysts for Change: Information Fluency, Web 2.0, Library 2.0, and the New Education Culture. N.p., 2007. Web. 18 Dec., 2008.

Maharana, Bulu, and Champeswar Mishra. "A Survey of Digital Information Literacy of Faculty at Sambalpur University." *Library Philosophy and Practice* 9.2 (2007). *E-prints in Lib. and Information Sciences.* Web. 18 Oct. 2008.

Martin, Allan. "Essential e-literacy." *Connected* 9(2003): 22–4. Web. 28 Oct. 2009.

Martin, Allan. "Literacies for the Digital Age." *Digital Literacies for Learning*.Eds. Allan Martin and Dan Madigan. London: Facet, 2006. 3–25. Web. 4 June. 2009.

Martin, Allan and Jan Grudziecki. "DigEuLit: Concepts and Tools for Digital Literacy Development." *ITALICS* 5.4 (2006): 249-67. *Directory of Open Access Journals*. Web. 4 April 2011.

McCarthy, Lesley. "Digital Literacy- Information Literacy- The New Literacies". *Proceedings ASET Conference: Untangling the Web: Establishing Learning Links. Melbourne, July 7-10, 2002*. Ed. S. McNamara and E. Stacey. N.p., 2002. Web. 18 Oct. 2008.

McShane, Ian. "Public Libraries, Digital Literacy and Participatory Culture." *Discourse: Studies in the Cultural Politics of Education* 32. 3 (2011): 383-97. *Routledge*. Web. 14 July 2012.

Media Awareness Network. Digital Literacy in Canada: From Inclusion to Transformation. A Submission to the Digital Economy Strategy Consultation. 2010. Web. 4 April 2011.

Nelson, Klara, Marcy Courier, and Gilbert W. Joseph "Teaching Tip: An Investigation of Digital Literacy Needs of Students." *Journal of Information Systems Education* 22.2 (2011): 95-109. *EBSCOhost*. Web. 20 Nov. 2012.

Owusu-Ansah, Edward K. "Information Literacy and the Academic Library: A Critical Look at a Concept and the Controversies Surrounding it." *Journal of Academic Librarianship* 29.4 (2003): 219-30. *ScienceDirect*. Web. 3 June 2011.

Prasher, R. G., and R. K. Sharma. ICT Based Information Management in Indian Libraries: Festschrift in Honour of Prof. (Dr.) K C Sahoo. New Delhi: Bookwell, 2012. Print.

Ray, Kathryn, and Joan Day. "Student Attitudes towards Electronic Information Resources." *Information Research* 4.2 (1998). *Google*. Web. 29 July 2009.

Rockman, H. "Successful Strategies for Integrating Information Literacy into the Curriculum." *Integrating Information Literacy into the Higher Education Curriculum: Practical Methods for Transformation*. Eds. Ilene F. Rockman et al. San Francisco, CA: Jossey-Bass, 2004. 47-91. Web. 18 Oct. 2010.

Romani, Cobo, and Cristobal Juan. Strategies to Promote the Development of E-competencies in the Next Generation of Professionals: European and International trends. Google. Web. 6 Feb 2013.

Rowlands, Ian, and David Bawden. "Digital Libraries: A Conceptual Framework." *Libri* 49.4 (1999): 192-202. Print.

Savolainen, R. "Network Competence and Information Seeking on the Internet: From Definitions towards a Social Cognitive Model." *Journal of Documentation* 58.2 (2002): 211–26. Print.

Sieberhagen, Elsabe Anee. The Design and Development of a Digital Information Literacy Programme for an Academic Library. Diss. U of So. Afr., 2005.N.p., 2005. U of So. Afr. Web. 18 Oct. 2008.

Society of College National and University Libraries. Information Skills in Higher Education: Briefing Paper, Prepared by the SCONUL Advisory Committee on Information Literacy. London: SCONUL, 1999. Web. 17 Mar. 2009.

Spitzer, Kathleen L., Michael B. Eisenberg, and Carrie A. Lowe. *Information Literacy: Essential Skills for the Information Age*. Santa Barbara, CA: Libraries Unlimited, 1998. Print.

Stubbings, Ruth, and Alan Brine. "Reviewing Electronic Information Literacy Training Packages." *ITALICS* 2.1 (2003). *Directory of Open Access Journals.* Web. 20 Jan. 2013.

Todd, Ross. "Information Literacy in Electronic Environments: Fantasies, Facts, and Futures." *Proceedings of the International Association of Scientific and Technological University Libraries Conferences* (2000). *Google.* Web. 18 Nov. 2009.

Tryon, Patricia J. Slagter van. "The Instructional Designer's Role in Forming University- Community Partnerships in Digital Literacy." *Tech Trends* 57.1 (2013): 52-8. *Springer Link.* Web. 16 Feb. 2013.

Welsh, Teresa S., and Melissa S Wright. *Information Literacy in Digital Age: An Evidence –Based Approach.* New Delhi: Chandos Publishing, 2010. Print.

Williams, Peter. "Opinion: Against Information Literacy." *Library and Information Update* 5.7–8 (2006): 20- 22. Web. 28 Aug. 2009.

Wilson, Myoung Chung. "To Dissect a Frog or Design an Elephant: Teaching Digital Information Literacy Through the Library Gateway". *63rd IFLA General Conference, Amsterdam, Aug. 31- Sep. 5, 1997. INSPEL* 32.3 (1998): 189-95. *Intl. Federation of Lib. Assn. and Inst.* Web. 18 Sept. 2008.

Zurkowski, Paul G. *The Information Service Environment: Relationships and Priorities.* Washington, DC: National Commission on Libraries and Information Science, 1974. *Google.* Web. 10 Jan. 2009.

■■■■

RESEARCH METHODOLOGY

2

2.1 Methodology and Data Collection Tools

It is imperative to decide upon the research methodology well in advance to carry out the research in a most effective and systematic way. According to Kothari, descriptive research also known as **Ex post facto research** in social sciences and in business studies which includes surveys and fact-finding enquiries (3). It involves the description, recording, analysis and interpretation of the existing conditions. It attempts to discover relationships between the existing non-manipulated variables (Best 25; Sharma 12). The main characteristic of this type of research is that the researcher has no control over the variables and can only report the present status. The method used in this type of research is **survey method**. Survey method is undertaken for making statistical inferences about the sample drawn from the population being studied. Further, the questionnaire method has been adopted to collect data in the present study.

Descriptive research designs with survey method was used with a fact-finding approach to carry out the current study of the DIL among the Master degree students and PhD scholars of the agricultural universities of Punjab, Haryana and Himachal Pradesh. The research methodology followed in this study involved collection and interpretation of data to find causal connections and relations for the existing status of DIL skills on sample of students.

2.2 Questionnaire Method

Questionnaire was invented by Sir Francis Galton and is considered as a significant research instrument. It consists of a series of questions for the purpose of gathering information from the respondents. Questio-nnaire method has an advantage over other survey methods because the respondents have sufficient time to read and answer the questions satisfactorily (Negi 156).

Keeping in view the scope, objectives and population of the present study as well as methodology of data collection adopted by the researchers in similar studies, questionnaire method was adopted as data collection tool in the present study.

2.3 Questionnaire Designing

According to Busha and Harter, a questionnaire must be developed carefully so that it accurately meets the objectives of research. It facilitates to obtain valid and reliable information so that specific hypotheses can be tested (61). A comprehensive structured questionnaire having closed-ended questions was designed, based on the objectives, to collect data. Questions were framed about the digital information requirements of the students, how effectively and efficiently students locate, evaluate and use information from digital resources of information (DRI). The questionnaire also focused on the ethical use of digital information by the students, the adoption of DIL in their social lives, their training needs with regard to DIL and role of the Library and Information Science (LIS) professionals in the information gathering process from DRI. It was divided into nine sections. **First section** contained questions regarding demographic profile of the respondents. Variables in demographic profile were age, gender, name of the university and class of study. The **second section** included questions on use of DRI by the respondents like purpose of using DRI, extent of DRI used, purpose of using e-mail and ethical use of DRI. **Third section** focused on accessing DRI and included questions on frequency of use of DRI, place of accessing DRI and quantum of information accessed from DRI. **Fourth section** focused on searching of information from DRI and comprised of questions relating to use of advanced search option, framing appropriate keywords, searching relevant information and other key factors in searching information from DRI. The **fifth section** was about student's ability to select authentic DRI. The questions regarding various problems faced while accessing DRI were asked in the **sixth section**. The **seventh section** included questions on the role of the library/university in using DRI like contents on use of DRI in the syllabus, training, efforts made and help rendered by the library staff. The **eighth section** aimed to assess the adoption of DIL in social lives of the respondents and ninth section included a question on suggestions to improve efficiency in using DRI.

2.4 Administration of Questionnaire

Best recommends that the questionnaires administered personally to the group of individuals has an advantage to establish rapport, to explain the purpose of study and meaning of terms that may not be clear (167). Thus, the questionnaires were personally distributed to the students who were contacted in libraries, hostels, play grounds, canteens and even classrooms after taking prior permission from the concerned teacher. During this process, a brief introduction to the importance of developing DIL skills in order to cope up with the ever expanding world of digital information was given to the respondents. They were also told about the significance of DIL in the academic & social life of an individual. Before filling the questionnaire, main objectives and impact of the research was explained to them. The respondents were given sufficient time to fill in the questionnaires so as to receive honest responses for the research.

2.5 Pre-testing of the Questionnaire

A pilot study was conducted on a small sample of ten respondents to pre-test the questionnaire. The questionnaires was personally distributed to five Master degree students and PhD scholars each to observe the problems if any faced by them while filling it with regard to various parameters like language, format, respondents' perception about terminology used and time taken to fill. They were personally asked to report any kind of problem which they faced while filling up the questionnaire. It took 10-15 minutes to fill the questionnaire by them. Minor modifications were made in the questionnaire as reported by them to make it clear and to obtain better research results.

2.6 Data Collection

Primary data was collected with the help of structured questionnaire (Annexure). Population consists of Master degree students and PhD scholars of Chaudhary Charan Singh Haryana Agricultural University, Hisar (HAU), Chaudhary Sarwan Kumar Himachal Pradesh Krishi Vishvavidyalaya, Palampur (HPKV) and Punjab Agricultural University, Ludhiana (PAU). Size of the population was 1227 students consisting of 901 Master degree students and 326 PhD scholars. According to Busha and Harter, survey method is characterized by selection of random sample from the population. The research is conducted on the sample and findings of the research are generalized to the larger group i.e. the population. Thus, sampling in survey research enables the researcher to gather information about target population without undertaking complete enumeration (54). Proportional stratified random sampling method was used in the present study to collect sample so as to give appropriate representation to both categories of students (Gupta 107). The population was divided into two strata viz. Master degree students and PhD scholars to achieve maximum homogeneity within the strata. Further, simple random sampling without replacement was used to draw sample from each stratum. The size of sample was 40% of the population i.e. 490 students, which is a satisfactory representative of the population. The data was collected during September 2010 to first week of February 2011.

2.7 Response Rate

The total number (no.) of 520 questionnaires was distributed amongst the Master degree students and PhD scholars of all the three subject universities, as detailed in Table 2.1 and 362 were received back duly filled in. Thus, the response rate was 69.6% with total sample size taken for this research as 362 respondents.

Table 2.1: Response Rate

Name of the university	No. of questionnaires distributed	No. of responses received
HAU	250	**184**
HPKV	70	**53**
PAU	200	**125**
Total	**520**	**362 (69.6%)**

2.8 Techniques Used in Analysis and Interpretation of Data

The data collected by filling up of the questionnaires was entered into MS-excel 2007 and a master data file was created which facilitated tabulation and analysis of data in desired form. The data was tabulated and analyzed using Minitab 16 for various statistical parameters as discussed in the following sub-sections. It was interpreted by means of relevant tables and charts (Taylor and Cihon).

2.8.1 Frequency distribution

The frequency distribution of data with respect to all the questions of the questionnaire having single as well as multiple option response was calculated and tabulated in the form of various graphs as well as tables. The frequency distribution of various questions with regard to important variables was also evaluated and presented in form of tables. Furthermore, the cross tabulation amongst the significant variables was calculated.

2.8.2 Percentage

The percentage of the data was calculated to compare the results and draw inferences. It was evaluated for all the questions which had single and multiple option response. The percentage was also evaluated with regard to important variables of various questions. It was also calculated for cross tabulations amongst various significant variables. The results were presented in form of graphs and tables.

2.8.3 Weighted index

Weighted index (W.I.) was used to analyze the data of the questions which involved ranking of the options by the respondents. W. I. is applied when all the items are not assigned equal importance or rank. It is an index where the important items are given more value than less important ones. Weight is assigned to each item relative to its importance and index number computed from these weights is called **weighted index**. The formula used to calculate W.I. is:

$$R_w = \frac{\sum_{i=1}^{n} R_i N_i}{\sum_{i=1} N_i}$$

Where;

N_i= No. of observation corresponding to i^{th} rank

R_i = n+1- i (n is the highest value rank and i^{th} is the actual rank value)

2.8.4 Karl Pearson's correlation coefficient

Correlation is a relationship that exists between two or more variables. Two variables are said to be correlated with each other if a change occurring in one of them causes a corresponding change in other. Karl Pearson's correlation coefficient (r) is a common measure of the correlation (linear dependence) between two variables X and Y, giving a value between +1 and −1 inclusive. A correlation of '+1' means that there is a **perfect positive linear relationship** between variables. A correlation of '-1' means that there is a perfect **negative linear relationship** between variables. A correlation of 0 means there is no linear relationship between the two variables (Gupta 258).

Mathematical Formula:

$$r = \frac{N\Sigma xy - (\Sigma x)(\Sigma y)}{\sqrt{[N\Sigma x^2 - (\Sigma x)^2][N\Sigma y^2 - (\Sigma y)^2]}}$$

Where:

N = number of pairs of scores

$\sum xy$ = sum of the products of paired scores

$\sum x$ = sum of x scores

$\sum y$ = sum of y scores

$\sum x^2$ = sum of squared x scores

$\sum y^2$ = sum of squared y scores

In the present study, the correlation between various scores namely digital information competency level score, social score, ethical use score and access score, as detailed below, was calculated using Karl Pearson's correlation coefficient.

Digital information competency level score

Digital information competency level score (DICLS) describes competency of an individual in evaluating and using digital information. In order to calculate this score the following questions at serial (sr.) numbers 25, 26, 27, 28, 30 and 31 of the questionnaire (Annexure), were selected:

25. In order to search information on 'Nutritional value of apple' from any of the CD-ROM databases subscribed by your university library, which is the most appropriate keyword out of the following?

    a) Apple-nutrition______ b) Apple and nutrition____

    c) Nutrition in apple____ d) Apple or nutrition___________

    e) Apple not nutrition_________

    f) Any other, please specify ____________

26. How much relevant information do you find while searching required information from digital resources of information?

a) All ______________ b) Almost all ____________

c) Little ____________ d) Very little ____________

e) None ____________

27. Searching information from digital resources of information is:

a) Very easy ___________ b) Easy ________________

c) Somewhat easy ______ d) Not easy______________

e) Difficult______________

28. On an average how many days do you usually take to search information on a given topic?

a) One day ___________ b) 2-5 days ____________

c) One week ____________ d) 8-10 days ____________

e) More than 10 days____________

30. Which of the following do you regard as most authentic resource of information for state-wise sex ratio in India?

a) www.india.gov.in _____________________________

b) www.cyberjournalist.org.in _____________________

c) www.infochangeindia.org _______________________

d) www.censusindia.gov.in ________________________

31. Which out of the following options is best resource to find out total number of agricultural universities and research institutes in India?

(i) www.icar.gov.in _____________________________

(ii) www.indiaedu.com ____________________________

(iii) www.cyberjournalist.org.in ____________________

(iv) www.career.webindia123.com __________________

The values as mentioned against each in the Table 2.2 were assigned to the questions chosen from the questionnaire (Annexure) to calculate DICLS. The questions no. 26, 27 and 28 were scored on a value ranging from '0' to '4' with value of '0' given to the least important and '4' given to most important attribute. On the other hand, the questions no. 25 and 31 having only one option as correct answer were assigned value'1' for the correct option and '0' for all the incorrect options. Similarly, the question no. 30 having more than one options as correct answers was assigned '2' for the most appropriate option followed by '1' for the second appropriate option and '0' for all incorrect options. Then the data which was collected from the respondents of this study was added with regard to these values, to give DICLS.

Table 2.2: Values Assigned to the Questions for Digital Information Competency Level Score

Sr. no. of the questionnaire	Options	Value assigned
25	Apple-nutrition	0
	Apple and nutrition	1
	Nutrition in apple	0
	Apple or nutrition	0
	Apple not nutrition	0
26	All	4
	Almost all	3
	Little	2
	Very little	1
	None	0
27	Very easy	4
	Easy	3
	Somewhat easy	2
	Not easy	1
	Difficult	0
28	One day	4
	2-5 days	3
	One week	2
	8-10 days	1
	More than 10 days	0
30	www.india.gov.in	1
	www.cyberjournalist.org.in	0
	www.infochangeindia.org	0
	www.censusindia.gov.in	2
31	www.icar.gov.in	1
	www.indiaedu.com	0
	www.cyberjournalist.org.in	0
	www.career.webindia123.com	0

Social score

Social score defines an individual's adoption of use of DRI in social lives. This score was calculated by assigning appropriate value to the options of the questions at serial numbers 39 to 43 of the questionnaire (Annexure). The questions are as mentioned below:

39. You get a chance to attend an international conference for example at National Academy of Agricultural Research Management, Hyderabad. How will you proceed?
    a) Immediately go to railway reservation counter and reserve a seat ______________________
    b) You would request your friend to immediately go to railway reservation counter and reserve a seat for yourself ______________________
    c) Check the reservation status from railway website first and then go to railway reservation counter to reserve a seat ______________________
    d) Check the reservation status from railway website first and then reserve a seat online ______________________
    e) Any other, please specify______________________

40. You come to know that today is the last day to deposit semester fee. What will you do?
    a) Withdraw money from bank using withdrawal form _________
    b) Withdraw money from bank using cheque ________________
    c) Withdraw money using ATM card ________________
    d) Request the bank staff for immediate help in withdrawal of money _________
    e) Any other, please specify______________________

41. Which format do you prefer to read the newspapers?
    a) Print_________
    b) Online__________
    c) Both__________

42. You wish to go abroad for further studies. How will you proceed? Please check all the answers which apply.

Sr.No.	Options used	Your Answer (√)
1	Go to immigration consultancy	
2	Consult your friends	
3	Consult your teachers	
4	Search for relevant information from the specific websites	
5	Any other, please specify________________	

43. In order to search suitable job for yourself, which of the following do you consult. Please check all the answers which apply.

Sr.No.	Options	Your Answer(√)
1	Employment news	
2	Newspapers	
3	College notice board	
4	Internet browsing	
5	Visiting the websites of the concerned companies	
6	Enquiring the placement cell of your institution	
7	Register online in various job sites	
8	Placement agency	
9	Any other option, please specify __________	

The Table 2.3 shows the values assigned to the questions 39 to 43 to calculate social score. The value '1' was assigned for the correct option and '0' for all the incorrect options. Similarly, the question with more than one appropriate option was assigned '2' for the most appropriate option followed by '1' for the second appropriate option and '0' for all incorrect options. Then the data was added with regard to these values, to calculate social score.

Table 2.3: Values Assigned to the Questions for Social Score

Sr. no. of the questionnaire	Options	Value assigned
39	Reservation counter	0
	Request a friend	0
	Check from website and reserve from reservation counter	1
	Check from website and reserve online	2
40	Withdrawal form	0
	Cheque	0
	ATM	1
	Request bank staff	0
41	Print	0
	Online	2
	Both	1
42	Immigration consultancy	0
	Consult friends	0
	Consult teachers	0
	Search relevant information from specific websites	1
43	Employment news	0
	Newspapers	0
	College notice board	0
	Internet	1
	Visiting the websites of the concerned companies	1
	Enquiring the placement cell of your institution	0
	Register online in various job sites	1
	Placement agency	0

Ethical use score

Ethical use score refers to tendency of ethical use and due acknowledgement given to the DRI used by the respondents. The ethical use score was calculated by assigning appropriate values to the question at serial numbers 12 and 13 in the questionnaire (Annexure 1). The questions are as mentioned below:

12. You have found that power-point presentation on your topic of credit seminar is available on Internet. What will you do? Please check all the answers which apply.

    a) Download it and use it___________

    b) Make changes in language and design and then use___________

    c) Use only idea/concept_____________

    d) Any other, please specify_________________

13. Do you cite the website or digital resource of information used by you? Please tick mark (√) the appropriate option.

    a) Always ____________ b) Almost everytime ____________

    c) Mostly_____________ d) Sometimes ________________

    e) Never______________

Table 2. 4 shows the values assigned to the questions 12 and 13 (Annexure) in order to calculate ethical use score. The question no. 12 was assigned values from '1' to '3' with value '1' given to the least important and '3' given to the most important attribute. On the other hand, question no. 13 was assigned values ranging from '0' to '4' with value of '0' given to the least important and '4' given to most important attribute. This was followed by adding the data with regard to these values, to calculate ethical use score.

Table 2.4: Values Assigned to the Questions for Ethical Use Score

Sr. no. of the questionnaire	Options	Value assigned
12	Download it and use it	1
	Make changes in language and design and then use	2
	Use only idea/concept	3
13	Always	4
	Almost every time	3
	Mostly	2
	Sometimes	1
	Never	0

Access score

Access Score shows frequency of access and quantum of information accessed from DRI. In order to calculate this score the following questions at serial numbers 16 and 21 of the questionnaire (Annexure), were selected:

16. How frequently do you access digital resources of information? Please tick mark (√) the appropriate option.

    a) Daily ____________ b) Once in two days ____________

    c) Weekly __________ d) Fortnightly __________________

    e) Monthly__________

    f) Any other, please specify___________________

21. The amount of information accessed by you from digital resources of information is (Please tick mark (√) the appropriate option):

    a) Adequate _______________ b) Inadequate ___________

    c) Very inadequate _______________________

The Table 2.5 gives the values assigned to the questions 16 and 21 in order to calculate access score. The question no. 16 was assigned value ranging from '1' to '5' with value of '1' given to the least important and '5' given to most important attribute. On the other hand, question number 21 was given value ranging from '0' to '2' with '0' given to the least important and '2' given to most important attribute. This was followed by adding the data with regard to these values, to calculate access score.

Table 2.5: Values Assigned to the Questions for Access Score

Sr. no. of the questionnaire	Options	Value assigned
16	Daily	5
	Once in two days	4
	Weekly	3
	Fortnightly	2
	Monthly	1
21	Adequate	2
	Inadequate	1
	Very inadequate	0

2.8.5 Pearson's Chi-squared Test

Pearson's Chi-squared test (χ^2) to determine two way associations of attributes is applied when there are two categorical variables from a single population, sampling method is simple random sampling, sample data are displayed in a contingency table and the expected frequency count for each cell of the table is at least 5. It is used to determine whether there is a statistically significant association between the two variables. The chi-square test provides a method for testing the association between the row and column variables in a two-way table (Gupta 324).

The formula for Pearson's Chi-squared test is:

$$\chi^2 = \frac{\sum_{i=1}^{n} (O_i - E_i)^2}{E_i}$$

Where

χ^2 = Pearson's cumulative test statistic, which asymptotically approaches a χ^2 distribution.

O_i = an observed frequency;

E_i = an expected (theoretical) frequency, asserted by the null hypothesis;

n = the number of cells in the table.

In the present study, the association of various scores namely DICLS, social score, ethical use score and access score (described in sub-sub-sections) was calculated with different variables of study namely age, gender, university and class; using Pearson's Chi-squared test.

REFERENCES

Best, John W. *Research in Education.* New Jersey: Prentice-Hall, 1959. Print.

Busha, Charles H, and Stephen P. Harter. *Research Methods in Librarianship: Techniques and Interpretation.* New York: Academic Press, 1980. Print.

Gupta, Santosh. *Research Methodology and Statistical Techniques.* New Delhi: Deep and Deep Pub, 2008. Print.

Kothari, C. R. *Research Methodology: Methods and Techniques.* 2nd ed. New Delhi: Wishwa Prakashan, 1990. Print.

Negi, M. S. *Research and Statistical Techniques in Library and Information Science.* New Delhi: Shree Pub and Distributors, 2005. Print.

Sharma, K. R. *Research Methodology.* New Delhi: National Pub House, 2002. Print.

Taylor, John K, and Cheryl Cihon. *Statistical Techniques for Data Analysis.* 2nd ed. Florida: Chapman and Hall/CRC, 2004. Print.

■■■■

USER CHARACTERSTICS 3

3.1 Introduction

It is very important to understand the structure and composition of the user community of an information system on the basis of their comparable characteristics so that they are served efficiently. According to Sridhar, No user-community of an information system is fully homogenous. Though a majority of the users of a system would come together for a particular purpose and are comparable by one or two criteria, they are divided among themselves by many individual characteristics (56). Keeping this in mind, the data collected to investigate the present problem of research with regard to demographic information of the respondents along with other aspects like purpose of using digital resources of information (DRI), ethical use of DRI and usefulness of DRI has been analyzed and interpreted in the current chapter. The purpose of this chapter is to describe the characteristic features of the population of this study, namely post-graduate students.

3.2 Post-graduate Students: General Background

The characteristics of users are numerous and there are several ways of grouping them. Apart from many characteristics and institutional environment it is the informational requirements of the users which has major impact on their DIL. The informational requirements of the Master degree students and PhD scholars in the Indian agricultural university differ on the basis of the nature and number of the credit courses offered to them. As per Punjab Agricultural University, there is a major difference in the number of major field courses and project research offered to master degree and PhD scholars (36-37). The number of major field courses for master degree students is much more than that of PhD and vice versa for project research (Table 3.1). This shows that course work for master degree students is more of classroom teaching whereas it is more research oriented for PhD scholars. Consequently, the informational requirements of master degree students is inclined more towards lecture notes, textbooks, assignments, etc whereas it is more of library oriented for PhD scholars because research forms the major component of their degree.

Table 3.1: Minimum Credit Hours required for Master Degree Students and PhD Scholars, PAU

Sr. no.		M.Sc/ M.Tech	MBA	MBA agribusiness	MCA	MJMC	PhD
1	Major field courses	20	53	55	60	27	12
2	Seminars	01	-	-	-	-	02
3	Supporting courses	05	06	07	06	08	05
4	Minor field courses	09	06	06	06	10	08
5	Research	25	12	12	18	15	45
Total		**60**	**77**	**80**	**90**	**60**	**72**

3.3 Libraries

It is appropriate to discuss in brief about the libraries of the three subject universities which are used by respective users for their informational requirements. The Nehru Library, HAU; University Library, HPKV and Mohinder Singh Randhawa Library, PAU form the hub of academic activities of these universities. These libraries cater to the informational needs of the respective faculty, students, researchers and extension specialists in modern perspective. Keeping pace with technological advancements these libraries serve their clientele with digital resources of information to enable expeditious, exhaustive, efficient, remote and multi-user access to its users. In addition to Online Public Access Catalogue and access to internet for academic purposes these libraries subscribed to three CDROM abstracting databases - AGRIS, AGRICOLA and various DRI also. The digital resources common in these libraries are CAB; two online databases- CeRA and Krishiprabha providing access to full text journals and theses respectively. Moreover, the library staff makes strenuous efforts to increase the use of digital resources and improve the DIL skills of the users by organizing users' orientation programmes and special training courses regularly. Library faculty teaches library and its services part from the compulsory course entitled 'Technical writing and Library and its services' to post-graduate students, in which emphasis is laid on efficient use of DRI thus improving their DIL skills.

3.4 Demographic Information

The first section of the questionnaire included the demographic information of the respondents viz. age group, gender, name of the university and class. The data so analyzed has been presented in graphical form in this section.

3.4.1 University and class-wise

The university-wise frequency distribution of the class i.e. Master degree students and PhD scholars is given in Fig. 3.1. The total respondents were 362 with 184 (50.8%) from HAU, 53 (14.6%) from HPKV and 125 (34.5%) from PAU. Figure 3.1 shows that out of total 362 respondents, there were 148 (40.8%) Master degree students and 36 (9.9%) PhD scholars were from HAU. In case of HPKV, there were 36 (9.9%) Master degree students and 17 (4.7%) PhD scholars. The number of respondents (R) in PAU was 76 (21%) Master degree students and 49 (13.5%) PhD scholars. In nut shell, there were a total of 260 (71.8%) Master degree students and 102 (28.2%) PhD scholars. The data was collected as per 'Proportional Stratified Random Sampling' method. The number of respondents of each university was corresponding to its population, with HAU having highest number of students (592) and thus, highest number of respondents (184).

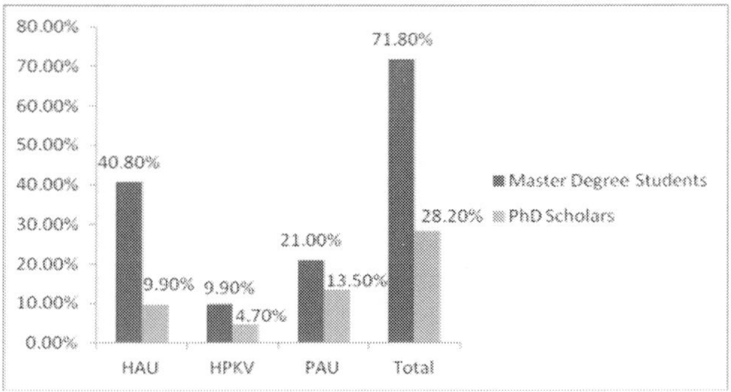

Note: Percentage is out of 362

Fig. 3.1: University-wise Distributions of Master degree students and PhD Scholars

3.4.2 University and gender-wise

The university and gender-wise distribution of the respondents, given in Fig. 3.2, included 195 (53.9%) male respondents and 167 (46.1%) female respondents. Further, out of 195 male respondents, 141 (39%) were from HAU, 19 (5%) from HPKV and 36 (9.9%) from PAU. The 167 female respondents included 43 (11.9%) from HAU, 35 (9.7%) from HPKV and 89 (24.6%) from PAU. The majority of respondents in HAU were male respondents whereas it constituted female respondents in HPKV and PAU.

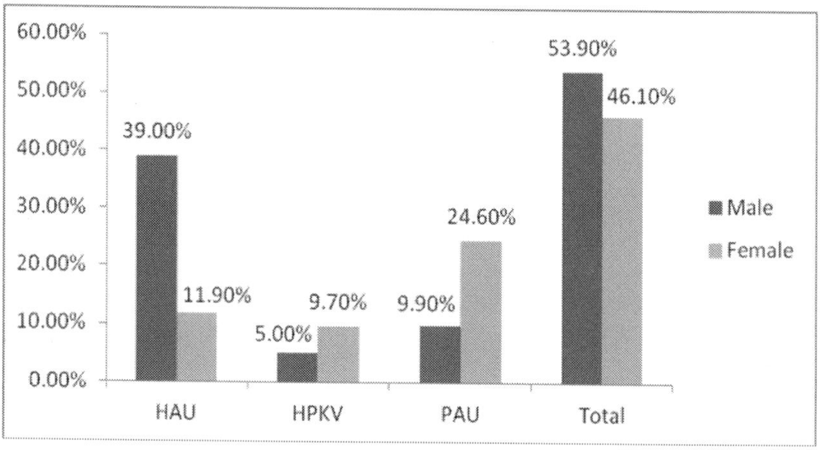

Note: Percentage is out of 362.

Fig. 3.2: University-wise Distributions of Male and Female Respondents

3.4.3 University and age-wise

The information regarding the age of respondent was sought in three different age groups as shown in Fig. 3.3. There were 180 (49.7%) respondents in the age group of 20-23 years, 151 (41.7%) in the age group of 24-27 years and 31 (8.6%) of more than 28 years of age. Out of these, there were 77 (21.3%) respondents from HAU, 33 (9.1%) from HPKV and 70 (19.3%) from PAU in the age group of 20-23 years. The age group of 24-27 years had 101 (27.9%) respondents from HAU, 13 (3.6%) from HPKV and 37 (10.2%) from PAU. And there were 6 (1.7%) respondents from HAU, 7 (1.9%) from HPKV and 18(5.1%) from PAU with age more than 28 years. The majority of respondents in HAU are from the age group 24-27 years whereas they are from the group 20-23 years in case of HPKV and PAU.

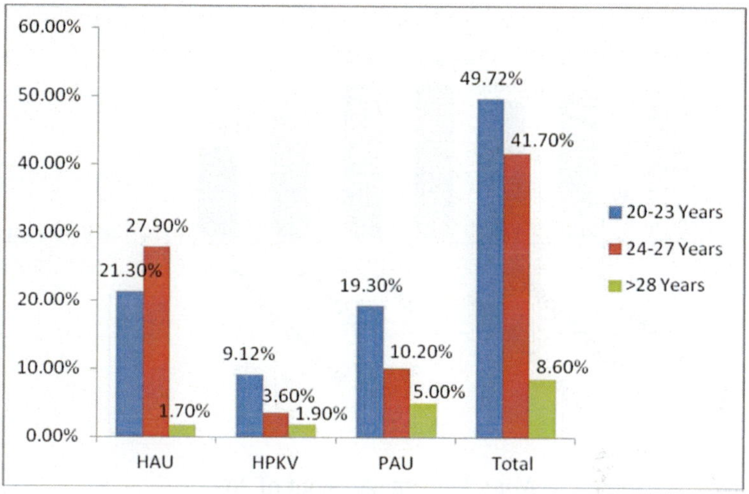

Note: Percentage is out of 362

Fig. 3.3: University-wise Distributions of Different Age Groups

3.5 Purpose of Using Digital Resources of Information

The data regarding purpose of using DRI was analyzed and frequency distribution is presented in different tables. The Fig. 3.4 shows various purposes for which the DRI were used. It was found that majority, 328 (90.6%) respondents used DRI to 'prepare for assignments, exams, etc'. 288 (79.6%) respondents used it for the purpose of 'thesis related work' followed by 287 (79.3%) for 'general awareness', 252 (69.6%) for 'writing research paper (for journal, conference, etc.)', and 222 (61.3%) for 'preparing class notes'. Only 162 (44.8%) of the respondents used DRI for 'doing some new experiments'.

The class-wise frequency distribution of data is shown in Table 3.2. It was found that 245 (94.2%) Master degree students used DRI for the purpose of 'preparing assignments, exams, etc.' whereas 97 (95.1%) PhD scholars used it for 'thesis related work'. The second purpose for which DRI were used was 'general awareness' by 208 (80%) Master degree students and 'writing research paper (for journal, conference, etc.)' by 94 (92.2%) PhD scholars. The reason for this trend might be that the basic objective of PhD scholars was doing research whereas in case of Master degree students though the research work is there but it is not so intense. It was also observed that DRI were used least for the purpose of 'doing new experiments' by both Master degree and PhD respondents.

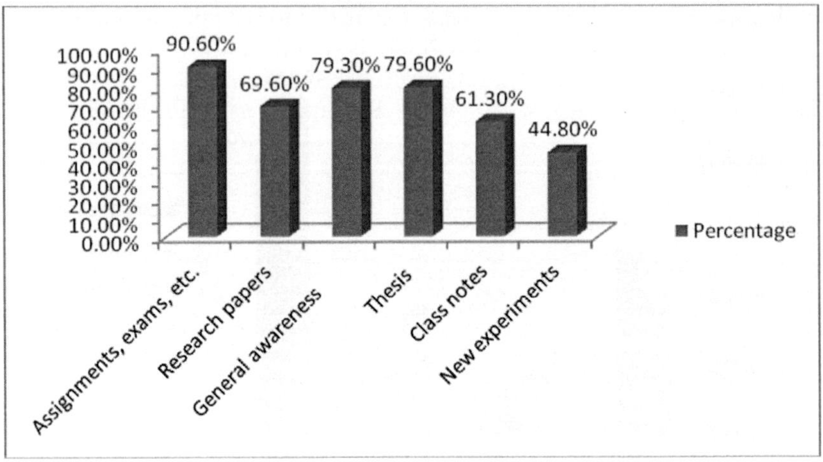

Note: Percentage is out of 362

Fig. 3.4: Purpose of Using Digital Resources of Information

Table 3.2: Purpose of Using Digital Resources of Information by Class

Purpose of using digital resources of information	Master degree students (% age out of 260)	PhD scholars (% age out of 102)
	R (%age)	R (%age)
Preparing for assignments, exams, etc.	**245 (94.2%)**	83 (81.4%)
Writing research paper (for journal, conference, etc.)	158 (60.8%)	94 (92.2%)
For general awareness	208 (80%)	79 (77.5%)
Thesis related work	191 (73.5%)	**97 (95.1%)**
Preparing class notes	148 (56.9%)	74 (72.6%)
Doing some new experiments	99 (38.1%)	63 (61.8%)

The university-wise frequency distribution of the respondents for the purpose of using DRI is given in Table 3.3. There were 92.4% respondents from HAU, 79.2% from HPKV and 92.8% from PAU who used DRI for the purpose of 'preparing assignments, exams, etc'. An equal number of respondents i.e., 79.2% from HPKV used DRI 'for general awareness'. The second purpose was 'thesis related work' for 84.2% respondents of HAU and 77.4% of HPKV whereas it was 'for general awareness' for 76.8% respondents of PAU. Again, it was observed that DRI were used least for the purpose of 'doing new experiments' by the respondents of all the three universities.

Table 3.3: Purpose of Using Digital Resources of Information by University

Purpose of using digital resources of information	HAU (% age out of 184)	HPKV (% age out of 53)	PAU (% age out of 125)
	R (%age)	R (%age)	R (%age)
Preparing for assignments, exams, etc.	**170 (92.4%)**	**42 (79.2%)**	**116 (92.8%)**
Writing research paper (for journal, conference, etc.)	137 (74.5%)	34 (64.2%)	81 (64.8%)
For general awareness	149 (81%)	**42 (79.2%)**	96 (76.8%)
Thesis related work	155 (84.2%)	41 (77.4%)	92 (73.6%)
Preparing class notes	113 (61.4%)	30 (56.6%)	79 (63.2%)
Doing some new experiments	79 (42.9%)	25 (47.2%)	58 (46.4%)

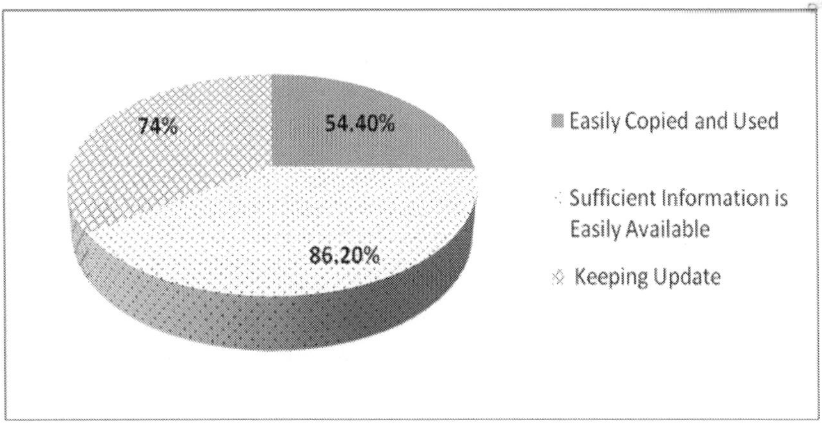

Note: Percentage is out of 362

Fig. 3.5: Digital Resources of Information are a Blessing

The respondents were asked about their opinion regarding the advantage of using DRI. The majority of respondents i.e. 312 (86.2%) opined that DRI are a blessing because 'sufficient amount of information is easily available on any given topic' (Fig. 3.5). It was followed by 268 (74%) respondents who opined that 'a person can keep update on all topics of interest' with the help of DRI. However, there were 197 (54.4%) respondents who regarded 'it can be easily copied and used'. It is clear that though majority of respondents are well versed of the advantages of DRI but still there is a considerable number who hold the attitude of unethical use of digital information.

3.6 Consultation of Digital References

Table 3.4 shows the percentage of digital references consulted by the respondents out of the total references. The frequency distribution and percentage of the data shows that majority of the respondents, 154 (42.5%), consulted 41%-60% digital references out of the total references. It was followed by 108 (29.8%) respondents who consulted only 21%-40% digital references, 52 (14.4%) consulted 61%-80% and 45 (12.4%) respondents consulted 1%-20% digital references. Very few respondents, only 3 (0.8%), consulted 81% -100% digital references. It is evident from the data shown in Table 3.4 that around 57% respondents consulted more than 40% digital references out of the total references. But still there were about 42% of respondents who consulted less than 40% digital references. Perhaps they may still prefer print media over online.

Table 3.4: Consultation of Digital References

Percentage of digital references consulted	No. of respondents	Percentage
1%-20%	45	12.4%
21%-40%	108	29.8%
41%-60%	**154**	**42.5%**
61%-80%	52	14.4%
81%-100%	3	0.8%

Note: Percentage is out of 362

Table 3.5 presents consultation of digital references by male and female respondents. It was found that majority of male respondents i.e. 44.1% and 40.7% female respondents consulted 41%-60% digital references. Moreover, it was also revealed that the consultation of digital references was more among male respondents as compared to female respondents.

Table 3.5: Consultation of Digital References by Gender

Percentage of digital references consulted	Male respondents (%age out of 195)	Female respondents (%age out of 167)
	R (%age)	R (%age)
1%-20%	29 (14.9%)	16 (9.6%)
21%-40%	53 (27.2%)	55 (32.9%)
41%-60%	**86 (44.1%)**	**68 (40.7%)**
61%-80%	26 (13.3%)	26 (15.6%)
81%-100%	1 (0.5%)	2 (1.2%)

The data was also analyzed university-wise and is presented in Table 3.6. It was found that majority 92 (50%) of respondents in HAU and 43 (34.4%) from PAU consulted 41%-60% digital references whereas 20 (37.7%) from HPKV consulted only 21%-40% of the same. The consultation of digital references was found to be least in HPKV.

Table 3.6: Consultation of Digital References by University

Percentage of digital references consulted	HAU (% age out of 184)	HPKV (% age out of 53)	PAU (% age out of 125)
	R (% age)	R (% age)	R (% age)
1%-20%	19 (10.3%)	11 (20.8%)	15 (12%)
21%-40%	57 (31%)	**20 (37.7%)**	31 (24.8%)
41%-60%	**92 (50%)**	19 (35.8%)	**43 (34.4%)**
61%-80%	16 (8.7%)	2 (3.8%)	34 (27.2%)
81%-100%	0	1 (1.9%)	2 (1.6%)

In order to study the trend of consulting digital references in Master degree students and PhD scholars, the data was analyzed class-wise and is presented in Table 3.7. It was found that 111 (42.7%) Master degree students and 43 (42.1%) PhD scholars, who constituted the major part of the respondents, were consulting 41%-60% digital references. Moreover, it was also revealed that the consultation of digital references by Master degree students was more than that of the PhD scholars which is in accordance with Crawford (33-34), who opined that the use of the internet decreases with year of study. Crawford conducted a study on the current students as well as alumni of the Glasgow Caledonian University to raise their awareness of the importance of IL and also to equip them with the skills to become independent users of information. The data was collected by means of electronic as well as paper questionnaire. The data was analyzed to realize that use of Internet declines gradually with the year of study. Likewise the present study also revealed that consultation of digital references was less in case of PhD scholars than Master degree students which are in confirmation to Crawford's study.

Table 3.7: Consultation of Digital References by Class

Percentage of digital references consulted	Master degree students (% age out of 260)	PhD scholars (% age out of 102)
	R (% age)	R (% age)
1%-20%	34 (13.1%)	11 (10.8%)
21%-40%	77 (29.6%)	31 (30.4%)
41%-60%	**111 (42.7%)**	**43 (42.1%)**
61%-80%	35 (13.4%)	17 (16.7%)
81%-100%	3 (1.2%)	0

Table 3.8: University and Class-wise Tabulation for Consultation of Digital References

Percent age of digital refere-nces consulted	HAU (184 R)		HPKV (53 R)		PAU (125 R)	
	Master degree students (%age out of 148)	PhD scholars (%age out of 36)	Master degree students (%age out of 36)	PhD scholars (%age out of 17)	Master degree students (%age out of 76)	PhD scholars (%age out of 49)
	R (%age)	R (%age)	R (%age)	R (%age)	R (%age)	R (%age)
1%-20%	16 (10.8 %)	3 (8.3 %)	9 (25%)	2 (11.8 %)	9 (11.8 %)	6 (12.2 %)
21%-40%	48 (32.5 %)	9 (25 %)	**14 (38.8 %)**	6 (35.2%)	15 (19.8 %)	**16 (32.7 %)**
41%-60%	**73 (49.3 %)**	**19 (52.8 %)**	11 (30.6 %)	**8 (47.1 %)**	**27 (35.5 %)**	**16 (32.7 %)**
61%-80%	11 (7.4 %)	5 (13.9 %)	1 (2.8 %)	1 (5.9 %)	23 (30.3 %)	11 (22.4 %)
81%-100%	0	0	1 (2.8 %)	0	(2.6 %)	0

The university and class-wise cross tabulation of data is presented in Table 3.8. It is evident that majority of Master degree students (49.3%) as well as PhD scholars (52.8%) of HAU, 47.1% PhD scholars of HPKV, 35.5% Master degree students and 32.7% PhD scholars from PAU consulted 41%-60% digital references. An equal number of 32.7% PhD scholars from PAU consulted 21%-40% digital references. Majority of Master degree students from HPKV i.e. 38.8% consulted only 21%-40% digital references. It is also evident that very few students i.e., three; consulted 81%-100% digital references as compared to 45 respondents who consulted only 1%-20%. However, it also revealed that, the consultation of digital references has decreased from Master degree students to PhD scholars incase of PAU which is in agreement with Crawford's study which states that Internet usage declines with year of study (33-34).

3.7 Purposc of Using E-mail

E-mail is a very popular use of internet and further internet itself is a very popular use of DRI. This study also explored the purpose(s) for which e-mail was used by the respondents. The overall trend as given in Fig. 3.6 shows that e-mail was used most for the purpose of 'communicating information/messages/images amongst friends' i.e. by 301 (83.1%) respondents, followed by 'receiving alerts of latest articles from various databases of full-text journals' by 250 (69.1%) respondents and 174 (48.1%) used it for the purpose of 'chatting'. It was found to be used least as 'discussion forum' only by 107 (29.6%) respondents. It is worth mentioning here that 69.1% respondents used e-mail to receive article alerts which is an academic use of e-mail. 83.1% of respondents used it for communicating and sharing information and 48.1% of respondents used it for the purpose of chatting. Both these purposes may be used for academic or social purpose but generally are used for social purpose. 29.6% used it as discussion forums which are characterized by sharing of information for academic purpose.

The same data was analyzed to explore the age-wise trend of using e-mail. The results given in Table 3.9, show that majority of respondents in all the three age groups; 81.7% in the age group of 20-23 years, 82.8% in 24-27 years and 93.5% of more than 28 years; used e-mail for 'communicating information/messages/images amongst friends' followed by for the purpose of 'receiving alerts of latest articles from various databases of full-text journals', 'chatting' and 'for discussion forums'. The data shows that use of e-mail for the purpose of communication increased with age. Perhaps the respondents of higher age groups are more focused towards research work, get more acquainted with use of technology and become more aware about the benefits of using DRI; thus, they use e-mail for communicating and sharing information, images, etc. more than the younger age groups.

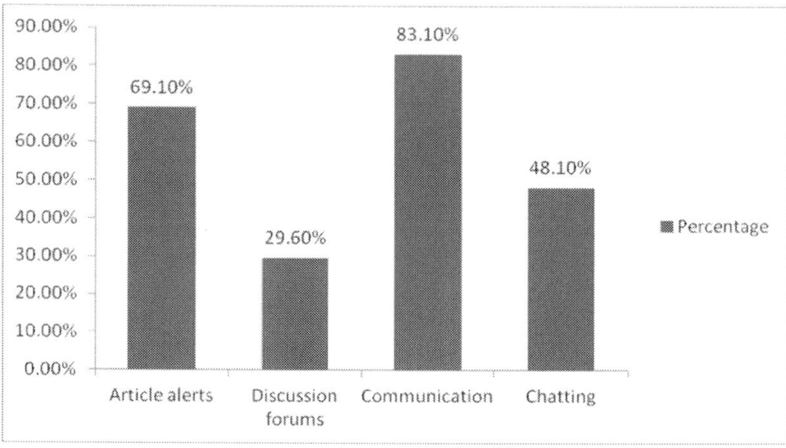

Note: Percentage is out of 362

Fig. 3.6: Purpose of Using E-mail

Table 3.9: Purpose of Using E-mail by Age

Purpose of using E-mail	20-23 Years (%age out of 180)	24-27 Years (%age out of 151)	>28 Years (%age out of 31)
	R (%age)	**R (%age)**	**R (%age)**
Receiving alerts of latest articles from various data-bases of full-text journals	110 (61.1%)	119 (78.8%)	21 (67.7%)
For discussion forums	56 (31.1%)	42 (27.8%)	9 (29%)
Communicating information/ messages/ images amongst friends	**147 (81.7%)**	**125 (82.8%)**	**29 (93.5%)**
Chatting	83 (46.1%)	76 (50.3%)	15 (48.4%)

The data was also analyzed gender-wise to investigate the trend of purpose of using e-mail amongst male and female respondents. Table 3.10 reveals majority of male (80%) and female (86.8%) respondents used e-mail for the purpose of 'communicating information/

messages/images amongst friends' followed by 'receiving alerts of latest articles from various databases of full-text journals', 'chatting' and 'for discussion forums'. The use of e-mails for the purpose of 'communicating information/messages/images amongst friends' was found more in female respondents as compared to male respondents. On the other hand it was used more by male respondents than female respondents for other purposes mentioned in the Table 3.10.

Table 3.10: Purpose of Using E-mail by Gender

Purpose of using E-mail	Male respondents (% age out of 195)	Female respondents (% age out of 167)
	R (%age)	R (%age)
Receiving alerts of latest articles from various databases of full-tcxt journals	140 (71.8%)	110 (65.9%)
For discussion forums	58 (29.7%)	49 (29.3%)
Communicating information/ messages/images amongst friends	**156 (80%)**	**145 (86.8%)**
Chatting	106 (54.4%)	68 (40.7%)

It is evident from Tables 3.9 and 3.10 that the majority of respondents from all the three age groups as well as both the sexes used e-mail for communication as compared to 'receiving e-mail alerts for latest articles published in journals'. The overall use of e-mail was found less for 'receiving e-mail alerts for latest articles published in journals' and 'discussion forums' as compared to 'communication' and 'chatting'. The reason might be that many students are still not aware of any such facility. It is thus, very important that the awareness regarding use of e-mail for 'receiving e-mail alerts for latest articles published in journals' and 'discussion forums' may also be increased by means of various users' education/ orientation programmes of the respective libraries.

3.8 Ethical Use of Digital Resources of Information

The issue of cyber plagiarism and unethical use of information is of great concern worldwide. Keeping in view the sensitivity of the issue, the understanding and awareness of the respondents regarding plagiarism was studied. The instance of use of relevant power point presentation available on internet was taken to study the ethical use of DRI. The overall response given in Fig. 3.7 shows that majority of respondents i.e., 257 (71%), would 'use only idea/ concept' of the relevant power point presentation available online. 75 (20.7%) respondents would 'make changes in language and design and then use it', however, 74 (20.4%) would simply 'download and use it'.

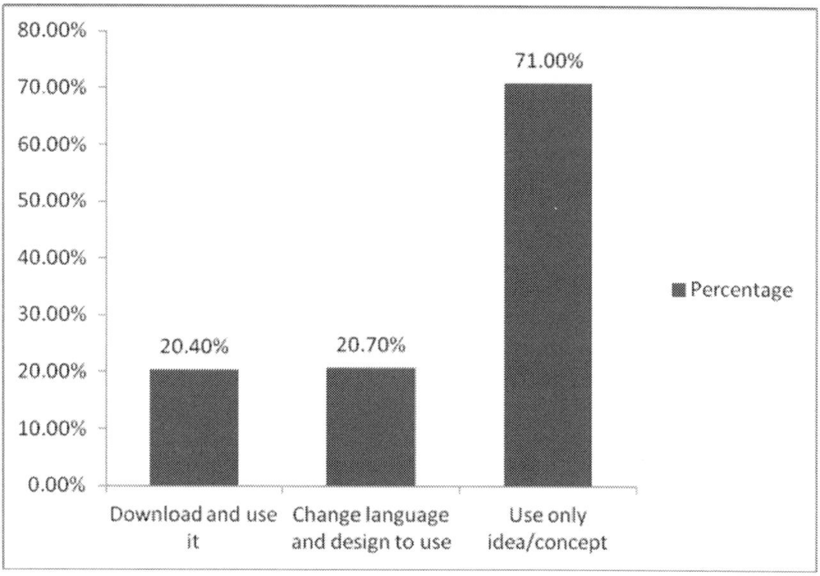

Note: Percentage is out of 362

Fig. 3.7: Use of PowerPoint Presentation

The data was analyzed for different age groups of respondents and is presented in Table 3.11. It was observed that majority of respondents in all the three age groups would 'use only idea/concept' of a relevant power point presentation available on internet. There were 128 (71.1%) respondents in age group of 20-23 years, 103 (68.2%) in age group of 24-27 years and 26 (83.9%) in the age group of more than 28 years for the same as shown in the Table 3.11.

Table 3.11: Use of Power Point Presentation by Age

Use of power point presentation	20-23 Years (%age out of 180)	24-27 Years (%age out of 151)	>28 Years (%age out of 31)
	R (%age)	R (%age)	R (%age)
Download and use it	32 (17.8%)	40 (26.5%)	2 (6.5%)
Make changes in language and design and then use it	40 (22.2%)	30 (19.9%)	5 (16.1%)
Use only idea/concept	**128 (71.1%)**	**103 (68.2%)**	**26 (83.9%)**

Similarly, the data was also analyzed gender-wise to find out the ethical use of DRI which is shown in Table 3.12. It was observed that majority of male respondents (64.1%) and female respondents (79%) would also 'use only idea/concept' of a relevant power point presentation available on internet. The female respondents were found to be more sensitive with regard to plagiarism as there were much more female respondents who would 'use only idea/concept'.

Table 3.12: Use of Power point Presentation by Gender

Use of power point presentation	Male respondents (%age out of 195)	Female respondents (%age out of 167)
	R (%age)	R (%age)
Download and use it	50 (25.6%)	24 (14.4%)
Make changes in language and design and then use it	40 (20.5%)	35 (21%)
Use only idea/concept	**125 (64.1%)**	**132 (79%)**

The data was also analyzed by university and class-wise to find the understanding of the respondents with regard to plagiarism of information from DRI. The Table 3.13 reveals that majority of Master degree students - 62.2% from HAU, 58.3% from HPKV and 72.4% from PAU as well as PhD scholars- 88.9% from HAU, 88.2% from HPKV and 85.7% from PAU have responded to 'use only idea/concept' even for a relevant powerpoint presentation available on internet. It was clear from the data that PhD scholars were more perceptive about the issue of plagiarism than Master degree students as the percentage of PhD scholars who would 'use only idea/concept' was more than Master degree students, may be because PhD scholars are more mature and responsible, moreover they understand the sanctity of the issue of ethical use of DRI as compared to Master degree students.

Table 3.13: University and Class-wise Tabulation of Use of PowerPoint Presentation

Use of powerpoint presentation	HAU (184 R)		HPKV (53 R)		PAU (125 R)	
	Master degree students (%age out of 148)	PhD scholars (%age out of 36)	Master degree students (%age out of 36)	PhD scholars (%age out of 17)	Master degree students (%age out of 76)	PhD scholars (%age out of 49)
	R (%age)	R (%age)	R (%age)	R (%age)	R (%age)	R (%age)
Download and use it	46 (31.1%)	2 (5.6%)	9 (25%)	0	9 (11.8%)	8 (16.3%)
Make changes in language and design and then use it	23 (15.5%)	6 (16.7%)	16 (44.4%)	3 (17.6%)	21 (27.6%)	6 (12.2%)
Use only idea/concept	**92 (62.2%)**	**32 (88.9%)**	**21 (58.3%)**	**15 (88.2%)**	**55 (72.4%)**	**42 (85.7%)**

It was clear from the Tables 3.11, 3.12 and 3.13 that majority of respondents according to age, gender, university and class used only idea/concept of any relevant DRI thus, majority of them have understanding of plagiarism. But still there were considerable number of respondents who would either directly download and use the DRI or would make minor changes and use it. The reason being was unconcerned attitude as well as lack of awareness about issue of plagiarism. Thus, the respondents need to be educated regarding the ethical use of DRI to improve their understanding about the issues of plagiarism. A case study conducted by Ercegovac (*What Students Say They Know, Feel, and Do about Cyber-Plagiarism and Academic Dishonesty? A Case Study*) identified various areas for which awareness and training programmes need to be designed on the basis of the weaknesses of the respondents and their levels of understanding with respect to ethical uses of information in general and digital networked resources in particular. Their understanding of the concept, attitudes and behaviors towards plagiarism in the web environment was studied. It was found that considerable numbers of respondents were ignorant about the concept of plagiarism. None of the respondents thought that plagiarism is copying someone's digital work and produce it under their own name. Therefore, it was recommended that awareness regarding ethical use of information may be created and efforts should be made with regard to different levels of understanding.

3.9 Citing Digital Resources of Information

The understanding of the respondents for ethical use of DRI was also explored by studying their responses for citing the DRI used by them. The overall response as given in Fig. 3.8 shows majority i.e. 141 (39%) of respondents 'mostly' cited the DRI consulted by them followed by 84 (23.2%) respondents who 'sometimes' cite, 82 (22.6%) would cite 'almost every time' and 54 (14.9%) would 'always' cite. It was revealed that only 1 (0.3%) of the respondents would 'never' cite the DRI consulted.

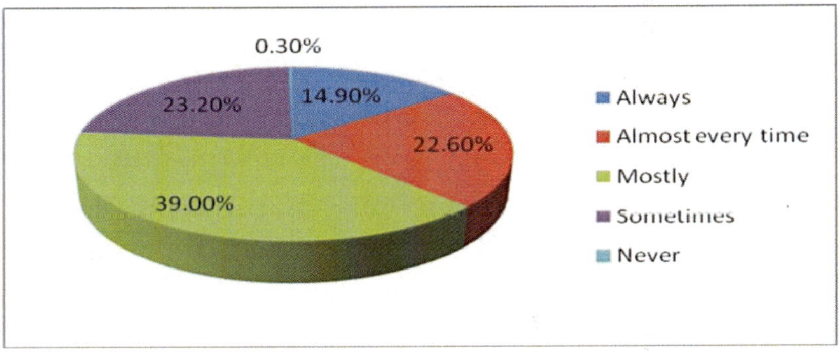

Note: Percentage is out of 362

Fig. 3.8: Citing Digital Resources of Information

Table 3.14: Citing Digital Resources of Information by Age

Citing digital resources of information	20-23 Years (%age out of 180)	24-27 Years (%age out of 151)	>28 Years (%age out of 31)
	R (%age)	**R (%age)**	**R (%age)**
Always	25 (13.9%)	21 (13.9%)	8 (25.8%)
Almost every time	34 (18.8%)	42 (27.8%)	6 (19.4%)
Mostly	**74 (41.1%)**	**55 (36.4%)**	**12 (38.7%)**
Sometimes	46 (25.6%)	33 (21.9%)	5 (16.1%)
Never	1 (0.6%)	0	0

The data was analyzed age-wise to find out the trend of citing DRI, see Table 3.14. It was observed that majority of respondents in all the three age groups- 41.1% in age group of 20-23 years, 36.4% in age group of 24-27 years and 38.7% in the age group of more than 28 years 'mostly' cited the DRI used by them. The practice of citing DRI was found to be least in the age group of 24-27 years.

Similarly, the data was also analyzed gender-wise to find the above information which is shown in Table 3.15. It was observed that majority of male respondents (34.9%) and female respondents (43.7%) 'mostly' cited the DRI used. The female respondents were found to be more responsive as compared to male respondents in citing the DRI used by them.

Table 3.15: Citing Digital Resources of Information by Gender

Citing digital resources of information	Male respondents (% age out of 195)	Female respondents (% age out of 167)
	R (%age)	R (%age)
Always	29 (14.9 %)	25 (15%)
Almost every time	44 (22.6%)	38 (22.7%)
Mostly	**68 (34.9%)**	**73 (43.7%)**
Sometimes	54 (27.7%)	30 (18%)
Never	0	1 (0.6%)

The data was analyzed by university and class-wise. The Table 3.16 reveals that majority of respondents- 37.8% Master degree students and 61.1% PhD scholars from HAU, 33.3% Master degree students from HPKV and 39.5% Master degree students as well as 36.7% PhD scholars from PAU 'mostly' cited the DRI used. However, an equal number of Master degree students of HPKV i.e., 33.3% 'sometimes' cited the DRI used. 29.4% of PhD scholars of HPKV cited the DRI used 'almost every time' and 'sometimes'.

It is also clear from the Tables 3.14, 3.15 and 3.16 that the number of respondents who cited the DRI consulted 'always' were less as compared to those who cite it 'almost every time' and 'mostly'. There was considerable number of respondents who 'sometimes' or 'never' cited the DRI consulted. Most of such respondents may not be conscious about the need and importance of citing the resource consulted which is in agreement with Ercegovac and Seamans. Ercegovac (*What Students Say They Know, Feel, and Do About Cyber-Plagiarism and Academic*

Dishonesty? A Case Study) found that 40% respondents failed to cite a document consulted. It was recommended that students must be educated regarding the ethical use of digital information. This study also recommended that the teachers should also be trained on this issue so that they are able to educate their students for citing the documents consulted. According to Seamans, respondents had little awareness about legal and ethical issues of electronic environment and paid less attention in evaluating information (*Student Perceptions of Information Literacy: Insights for Librarians* 112-23). So, it is concluded that there was dearth of education amongst the respondents regarding ethical use of DRI which can be improved by proper awareness amongst them.

Table 3.16: University and Class-wise Tabulation of Citing Digital Resources of Information

Citing digital resources of information	**HAU**		**HPKV**		**PAU**	
	Master degree students (%age out of 148)	**PhD scholars (%age out of 36)**	**Master degree students (%age out of 36)**	**PhD scholars (%age out of 17)**	**Master degree students (%age out of 76)**	**PhD scholars (%age out of 49)**
	R (%age)	**R (%age)**	**R (%age)**	**R (%age)**	**R (%age)**	**R (%age)**
Always	21 (14.2%)	8 (22.3%)	6 (16.7%)	4 (23.5%)	2 (2.6%)	13 (26.5%)
Almost every time	36 (24.3%)	3 (8.3%)	6 (16.7%)	**5 (29.4%)**	21 (27.6%)	11 (22.5%)
Mostly	**56 (37.8%)**	**22 (61.1%)**	**12 (33.3%)**	3 (17.6%)	**30 (39.5%)**	**18 (36.7%)**
Sometimes	35 (23.7%)	3 (8.3%)	**12 (33.3%)**	**5 (29.4%)**	22 (29%)	7 (14.3%)
Never	0	0	0	0	1 (1.3%)	0

3.10 Usefulness of Digital Resources of Information

The usefulness of various DRI available in all the three agricultural universities by the respondents was calculated using W.I. It is evident from Table 3.17 that Consortium for e-Resources in Agriculture (CeRA) was the most useful DRI as it got first rank followed by Commonwealth Agricultural Bureaux (CAB) which got second rank. 'Other digital resources available through Internet' got third rank, Krishiprabha got fourth rank, AGRIS got fifth rank and AGRICOLA got sixth rank.

CeRA was found to be most useful database as it is a consortium for e-resources in agriculture sponsored by Indian Council of Agricultural Research (ICAR). This database provides descriptive, current and statistical information by providing access to almost all the core full-text online journals in the field of agricultural sciences and allied areas at a single platform and is campus wide available in all the Indian state agricultural universities and agricultural research institutes of ICAR. In addition to full-text journals it also provides the facility of inter-library loan, under which the copies of the articles of those journals which are not covered under CeRA, but are available in any of the member libraries, is made available to the applicant. CAB, AGRIS and AGRICOLA are abstracting databases in the field of agricultural sciences and allied areas. These databases provide access to current, bibliographic and statistical information through the abstracts along with complete bibliographic details of the latest ongoing research worldwide. Krishiprabha is an online database of full-text PhD theses submitted from the year 2000 onwards in all the agricultural universities and research institutes of ICAR. It provides conceptual, descriptive and statistical information on all the disciplines in the field of agriculture and allied areas. 'Other digital resources of information available through the internet' refers various open access DRI in the field of agriculture and allied areas for example website of Food and Agricultural Organization, World Health Organization, Agriculture Network Information Centre, International Database on Insect Disin-festations and Sterilization, etc. Such DRI provides access to current, statistical, bibliographic, descriptive and conceptual information.

Table 3.17: Usefulness of Digital Resources of Information

Digital resources of information	Usefulness of digital resources of information					W.I.	Rank as in W.I.
	4	3	2	1	0		
	Very useful (R)	Useful (R)	Satisfactory (R)	Of little use (R)	Of no use (R)		
CeRA	207	115	21	11	8	3.39	1
CAB	151	102	67	26	16	2.96	2
Other DRI available through the Internet	95	146	76	32	13	2.77	3
Krishiprabha	86	102	70	68	36	2.37	4
AGRIS	77	90	94	46	55	2.24	5
AGRICOLA	79	94	75	53	61	2.21	6

3.11 Use of Digital Resources of Information for Type of Information

The DRI available in all the three agricultural universities as mentioned in Table 3.18 can be used for accessing descriptive, statistical, current, conceptual and bibliographic information by the Master degree students and PhD scholars. The type of information for which these DRI were used by Master degree students and PhD scholars was calculated and is given in Table 3.18. The analysis of data reveals that majority of the respondents- 35.1%, 26.5% and 39.5% use AGRIS, AGRICOLA and CAB respectively for bibliographic information. CeRA and Krishi-prabha were used by 56.4% and 24.3% respondents respectively, for descriptive information. 'Other DRI available through internet' were used by 42.3% respondents for accessing current information. The type of information actually available from all these DRI has already been discussed in section 3.9 of this chapter.

Table 3.18: Using Digital Resources of Information for Type of Information

Digital resources of information	Descriptive infor-mation	Statistical infor-mation	Current infor-mation	Conce-ptual infor-mation	Biblio-graphic infor-mation
	R (%age)	R (%age)	R (%age)	R (%age)	R (%age)
AGRICOLA	101 (27.9%)	42 (11.6%)	73 (20.2%)	23 (6.4%)	**127 (35.1%)**
AGRIS	69 (19.1%)	29 (8%)	53 (14.6%)	28 (7.7%)	**96 (26.5%)**
CAB	107 (29.6%)	30 (8.3%)	76 (21%)	51 (14.1%)	**143 (39.5%)**
CeRA	**204 (56.4%)**	55 (15.2%)	153 (42.3%)	112 (31%)	62 (17.1%)
Krishiprabha	**88 (24.3%)**	33 (9.1%)	44 (12.2%)	60 (16.6%)	68 (18.8%)
Other DRI available through the Internet	95 (26.2%)	74 (20.4%)	**153 (42.3%)**	86 (23.8%)	37 (10.2%)

Note: Percentage is out of 362

The data was also analyzed university and class-wise to find out the use of DRI for specific information. The results for HAU are shown in Table 3.19. It was found that majority of Master degree students 52% used AGRICOLA, 35.1% used AGRIS and 47.3% used CAB for accessing bibliographic information. 66.2% used CeRA and 28.4% used

Krishiprabha for descriptive information. 37.8% used 'other DRI available through internet' for accessing current information. On the other hand, for PhD scholars it was observed that 61.1% and 41.7% used AGRICOLA and AGRIS respectively for descriptive information whereas 50% used CAB for bibliographic information. 58.3% used CeRA again for descriptive information. 33.3% used Krishiprabha for current information and 33.3% used 'other DRI available through Internet' for accessing statistical information.

Table 3.19: Using Digital Resources of Information for Type of Information Class-wise in HAU

Digital resources of information	Descriptive information		Statistical information		Current information		Conceptual information		Bibliographic information	
	Master degree students (%age out of 148)	PhD scholars (%age out of 36)	Master degree students (%age out of 148)	PhD scholars (%age out of 36)	Master degree students (%age out of 148)	PhD scholars (%age out of 36)	Master degree students (%age out of 148)	PhD scholars (%age out of 36)	Master degree students (%age out of 148)	PhD scholars (%age out of 36)
	R (%age)	R (%age)	R (%age)	R (%age)	R (%age)	R (%age)	R (%age)	R (%age)	R (%age)	R (%age)
AGRICOLA	48 (32.4%)	**22 (61.1%)**	16 (10.8%)	7 (19.4%)	23 (15.5%)	6 (16.7%)	11 (7.4%)	0	**77 (52%)**	17 (47.2%)
AGRIS	32 (21.6%)	**15 (41.7%)**	11 (7.4%)	5 (13.9%)	19 (12.8%)	6 (16.7%)	3 (2%)	2 (5.6%)	**52 (35.1%)**	12 (33.3%)
CAB	48 (32.4%)	15 (41.7%)	10 (6.8%)	7 (19.4%)	26 (17.6%)	3 (8.3%)	22 (14.9%)	3 (8.3%)	**70 (47.3%)**	**18 (50%)**
CeRA	**98 (66.2%)**	**21 (58.3%**	27 (18.3%)	5 (13.9%)	83 (56.1%)	10 (27.8%)	61 (41.2%)	7 (19.4%)	26 (17.6%)	7 (19.4%)
Krishiprabha	**42 (28.4%)**	6 (16.7%)	18 (12.2%)	3 (8.3%)	12 (8.1%)	**12 (33.3%)**	33 (22.3%)	7 (19.4%)	40 (27%)	3 (8.3%)
Other DRI available through the Internet	26 (17.6%)	6 (16.7%)	24 (16.2%)	**12 (33.3%)**	**56 (37.8%)**	8 (22.2%)	45 (30.4%)	5 (13.9%)	13 (8.8%)	2 (5.6%)

The type of information for which the DRI are used by HPKV students is shown in Table 3.20. The majority of Master degree students 38.9%, 52.8% and 38.9% used AGRCOLA, CAB and 'other DRI available through Internet' for accessing current information. 44.4% and 33.3% used CeRA and Krishiprabha for descriptive information. 22.2% used AGRIS for accessing bibliographic information. On the other hand, for PhD scholars it was observed that 35.3% used AGRICOLA and CAB, 41.2% used CeRA and 23.5% used Krishiprabha respectively for descriptive information whereas 29.4% and 41.2% used AGRIS and 'other DRI available through Internet' for current information. 35.3% also used CAB for accessing bibliographic information.

Table 3.20: Using Digital Resources of Information for Type of Information Class-wise in HPKV

Digital resou-rces of infor-mation	Descriptive information		Statistical information		Current information		Conceptual information		Bibliographic information	
	Master degree students (%age out of 36)	PhD scho-lars (%age out of 17)	Master degree students (%age out of 36)	PhD scho-lars (%age out of 17)	Master degree students (%age out of 36)	PhD scholars (%age out of 17)	Master degree students (%age out of 36)	PhD scho-lars (%age out of 17)	Master degree students (%age out of 36)	PhD scho-lars (%age out of 17)
	R (%age)	R (%age)	R (%age)	R (%age)	R (%age)	R (%age)	R (%age)	R (%age)	R (%age)	R (%age)
AGRI-COLA	8 (22.2%)	**6 (35.3%**	2 (5.6%)	1 (5.9%)	**14 (38.9%)**	3 (17.6%)	2 (5.6%)	2 (11.8%)	7 (19.4%)	4 (23.5%)
AGRIS	1 (2.8%)	4 (23.5%)	3 (8.3%)	0	7 (19.4%)	**5 (29.4%)**	4 (11.1%)	1 (5.9%)	**8 (22.2%)**	4 (23.5%)
CAB	11 (30.6%)	**6 35.3%)**	1 (2.8%)	0	**19 (52.8%)**	5 (29.4%)	1 (2.8%)	3 (17.6%)	3 (8.3%)	**6 35.3%)**
CeRA	**16 (44.4%)**	**7 (41.2%**	2 (5.6%)	3 (17.6%)	15 (41.7%)	2 (11.8%)	9 (25%)	2 (11.8%)	1 (2.8%)	3 (17.6%)
Krishi-prabha	**12 (33.3%)**	**4 (23.5%**	4 (11.1%)	2 (11.8%)	4 (11.1%)	2 (11.8%)	1 (2.8%)	3 17.6%)	7 (19.4%)	1 (5.9%)
Other DRI available through the Internet	11 (30.6%)	6 (35.3%)	12 (33.3%)	4 (23.5%)	**14 (38.9%)**	**7 (41.2%)**	5 (13.9%)	2 (11.8%)	4 (11.1%)	2 (11.8%)

The results to find out the use of DRI for the type of information in PAU are given in Table 3.21. It was found that 19.7% and 61.8% Master degree students used AGRICOLA and 'other DRI available through Internet' respectively for accessing current information. 17.1% used AGRIS to access conceptual information. 38.2% used CAB for accessing bibliographic information. 43.4% used CeRA and 21.1% used Krishiprabha for descriptive information. On the other hand, 24.5% and 42.9% PhD scholars used AGRICOLA and 'other DRI available through Internet' for accessing current information. 34.7% used CAB for bibliographic information. 20.4%, 59.2% and 16.3% used AGRIS, CeRA and Krishiprabha respectively for descriptive information.

Table 3.21: Using Digital Resources of Information for Type of Information Class-wise in PAU

Digital resources of information	Descriptive information		Statistical information		Current information		Conceptual information		Bibliographic information	
	Master degree students (%age out of 76)	PhD scholars (%age out of 49)	Master degree students (%age out of 76)	PhD scholars (%age out of 49)	Master degree students (%age out of 76)	PhD scholars (%age out of 49)	Master degree students (%age out of 76)	PhD scholars (%age out of 49)	Master degree students (%age out of 76)	PhD scholars (%age out of 49)
	R (%age)	R (%age)	R (%age)	R (%age)	R (%age)	R (%age)	R (%age)	R (%age)	R (%age)	R (%age)
AGRICOLA	7 (9.2%)	10 (20.4%)	12 (15.8%)	4 (8.2%)	**15 (19.7%)**	**12 (24.5%)**	4 (5.3%)	4 (8.2%)	13 (17.1%)	7 (14.3%)
AGRIS	7 (9.2%)	**10 (20.4%)**	5 (6.6%)	5 (10.2%)	10 (13.2%)	6 (12.2%)	**13 (17.1%)**	5 (10.2%)	11 (14.5%)	9 (18.4%)
CAB	13 (17.1%)	14 (28.6%)	9 (11.8%)	3 (6.2%)	12 (15.8%)	11 (22.5%)	13 (17.1%)	9 (18.4%)	**29 (38.2%)**	**17 (34.7%)**
CeRA	**33 (43.4%)**	**29 (59.2%)**	15 (19.7%)	3 (6.2%)	27 (35.5%)	16 (32.7%)	22 (29%)	11 (22.5%)	13 (17.1%)	12 (24.5%)
Krishiprabha	**16 (21.1%)**	**8 (16.3%)**	4 (5.3%)	2 (4.1%)	10 (13.2%)	4 (8.2%)	11 (14.5%)	5 (10.2%)	11 (14.5%)	6 (12.2%)
Other DRI available through the Internet	35 (46.1%)	11 (22.5%)	13 (17.1%)	9 (18.4%)	**47 (61.8%)**	**21 (42.9%)**	19 (25%)	10 (20.4%)	9 (11.8%)	7 (14.3%)

Tables 3.17 to 3.21 show that CeRA is the most preferred DRI which is used for accessing descriptive information by the Master degree students as well as PhD scholars of all the three universities. CAB is second most preferred DRI and is used to access bibliographic information by both Master degree students and PhD scholars of all the three universities except for Master degree students of HPKV. As already mentioned in section 3.9, CeRA provides campus wide access to almost all the core journals in the field of agriculture and allied areas to all the agricultural universities and agricultural research institutes of India. That's why it is mostly used DRI for accessing descriptive information. CAB on the other hand is an abstracting database in the field of agricultural sciences and allied areas, published by Commonwealth Agricultural Bureaux International. It provides access to the abstracts along with complete bibliographic details of the latest ongoing research worldwide. That's why it is used for accessing references for the purpose of their literature review by the researchers.

3.12 Conclusion

In nut shell, it was found that majority of respondents used DRI to 'prepare for assignments, exams, etc' and believed that using DRI is a blessing because 'sufficient amount of information is easily available on any given topic'. The consultation of digital references was 51% -75% out of the total references, which is not satisfactory. E-mail was used most for the purpose of communication. It was also found that majority of respondents were aware of plagiarism as they would 'use only idea/concept' of a relevant powerpoint presentation available on internet. However, number of respondents who would 'always' cite the DRI consulted was very less as compared to number of respondents who would cite it 'sometimes'. Moreover, there was considerable number of respondents who believed that digital information 'can be easily copied and used'. Thus, the respondents need to be educated regarding the ethical use of DRI. Furthermore, the type of information for which the online and CDROM databases were used revealed that AGRICOLA, AGRIS and CAB CDROM databases were used by the majority for 'bibliographic information'; CeRA and Krishiprabha for 'descriptive information' and 'other digital resources available through Internet' for 'current information'. CeRA was found to be most useful out of all the databases available in the three universities. CAB was second most useful database.

After analyzing the respondents for various aspects like purpose of using DRI, ethical use of DRI and usefulness of DRI; other aspects like their capability in accessing DRI, searching information from DRI and selecting authentic DRI needs to be explored further.

REFERENCES

Crawford, John. "The Use of Electronic Information Services and Information Literacy: A Glasgow Caledonian University Study." *Journal of Librarianship and Information Sciences* 38.1 (2006): 33-44. Print.

Ercegovac, Zorana. "What Students Say They Know, Feel, and Do About Cyber-Plagiarism and Academic Dishonesty? A Case Study." *Proceedings of the American Society for Information Science and Technology* 42.1 (2005). *Wiley Interscience*. Web. 18 Nov. 2009.

Punjab Agricultural University. *Resident Instruction Bulletin.* Ludhiana: Punjab Agricultural University, 2014. Print.

Seamans, Nancy H. "Student Perceptions of Information Literacy: Insights for Librarians." *Reference Services Review* 30.2 (2002): 112-23. *Emerald.* Web. 29 Oct. 2009.

Sridhar, M. S. Information Behaviour of Scientists and Engineers: A Case Study of Indian Space Technologists. New Delhi: Concept Publishing Co., 1995. Print.

■■■■

ACCESSING, SEARCHING AND SELECTING DIGITAL RESOURCES OF INFORMATION

4

4.1 Introduction

The digital information literacy (DIL) skills of the students in the present study were assessed by analyzing the data using MS Excel 2007 and Minitab 16 for frequency distribution, percentage, weighted index (W.I.), correlation and association of attributes. This chapter covers the analyses and discussion of data for various aspects like accessing digital resources of information (DRI), searching information from DRI and selecting authentic DRI.

4.2 Accessing Digital Resources of Information

This section covers various aspects of accessing DRI like frequency of access, place of access, adequacy of information accessed, mode of access and acquisition of information from DRI. The frequency and percentage of data was calculated to study these aspects, which is discussed in the following sub-sections.

4.2.1 Frequency of accessing digital resources of information

The overall response showing the frequency of accessing DRI is given in Figure 4.1. It was found that 119 (32.9%) respondents accessed the DRI 'once in two days' followed by 112 (30.9%) who used it 'weekly', 107 (29.6%) accessed it 'daily' and 17 (4.7%) accessed it fortnightly. There were very few i.e., 7 (1.9%) who accessed it on 'monthly' basis.

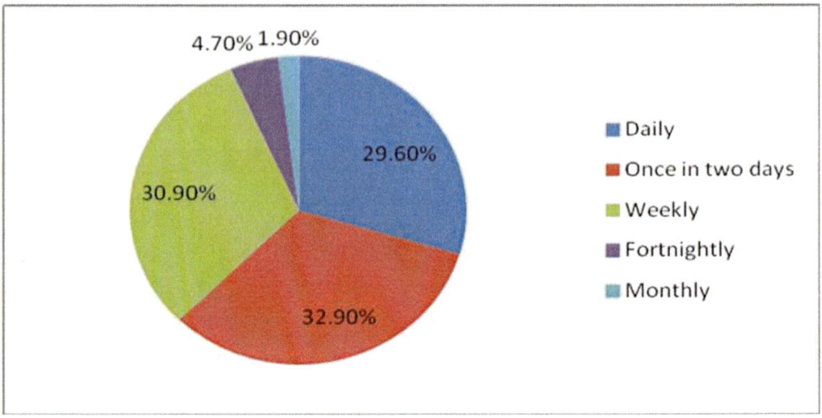

Note: Percentage is out of 362

Fig. 4.1: Frequency of Accessing Digital Resources of Information

The university-wise frequency distribution and percentage of data as shown in Table 4.1 shows that majority of respondents (40.8%) in HAU access DRI 'once in two days', 50.9% respondents in HPKV access the DRI 'weekly' whereas 44.8% of respondents in PAU access 'daily'. It was found that the frequency of accessing of DRI was maximum in PAU.

Table 4.1: Frequency of Accessing Digital Resources of Information University-wise

Frequency of accessing digital resources of information	**HAU (%age out of 184)**	**HPKV (%age out of 53)**	**PAU (%age out of 125)**
	R (%age)	**R (%age)**	**R (%age)**
Daily	41 (22.3%)	10 (18.9%)	**56 (44.8%)**
Once in two days	**75 (40.8%)**	13 (24.5%)	31 (24.8%)
Weekly	56 (30.4%)	**27 (50.9%)**	29 (23.2%)
Fortnightly	10 (5.4%)	1 (1.9%)	6 (4.8%)
Monthly	2 (1.1%)	2 (3.8%)	3 (2.4%)

The Table 4.2 shows the class-wise frequency distribution of data with regard to frequency of accessing DRI. Number of respondents (R) which constituted major part in case of Master degree students was 35% who accessed DRI 'weekly'. Majority of PhD scholars i.e. 39.2% accessed DRI 'daily'. The frequency of accessing DRI by PhD scholars was more than Master degree students. The research work is more intense in case of PhD scholars thus; they may access DRI and use the library constantly.

Table 4.2: Frequency of Accessing Digital Resources of Information Class-wise

Frequency of accessing digital resources of information	Master degree students (%age out of 260)	PhD scholars (%age out of 102)
	R (%age)	R (%age)
Daily	67 (25.8%)	**40 (39.2%)**
Once in two days	83 (31.8%)	36 (35.3%)
Weekly	**91 (35%)**	21 (20.6%)
Fortnightly	16 (6.2%)	1 (1%)
Monthly	3 (1.2%)	4 (3.9%)

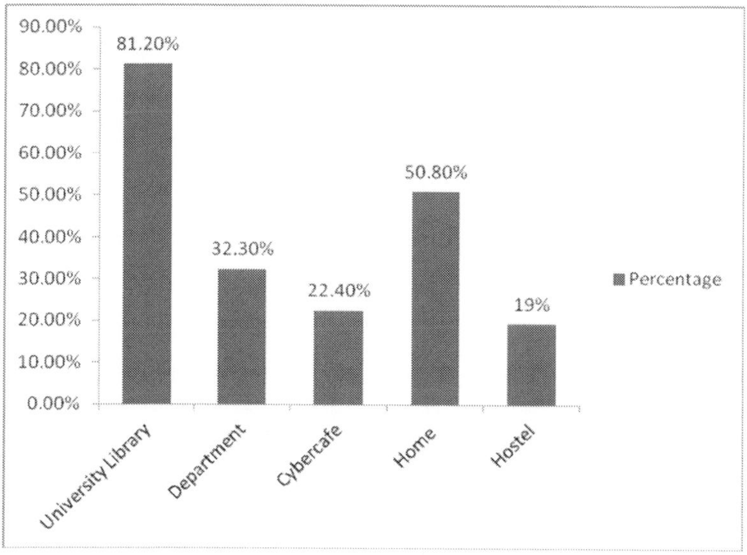

Note: Percentage is out of 362

Fig. 4.2: Place of Accessing Digital Resources of Information

4.2.2 Place of accessing digital resources of information

The response in general regarding the place of accessing DRI is given in Fig. 4.2. It was found that by and large 294 (81.2%) respondents preferred 'university library' followed by 184 (50.8%) who preferred to access DRI from their 'homes', 117 (32.3%) from their 'departments', 81 (22.4%) from 'cybercafé' and 70 (19%) from their 'hostels'. The university library is a hub of all the academic activities as well as the central place of the university. The students might feel it conveniently approachable for accessing DRI.

Table 4.3 shows age-wise frequencies and percentage of data regarding place of accessing DRI. Majority of respondents in the age group of 20-23 years (76.7%) and 24-27 years (89.4%) accessed DRI from 'university library', however, 71% of age more than 28 years accessed it from their 'department'. The younger age groups may need help from the library staff for accessing DRI whereas the age group of more than 28 years being more experienced is possibly able to access DRI of their own from their departments. The second preference for the accessing the DRI was found to be 'home' by both 20-23 years and 24-27 years age group whereas it was 'university library' by the respondents of more than 28 years age.

Table 4.3: Place of Accessing Digital Resources of Information Age-wise

Place of accessing digital resources of information	20-23 Years (%age out of 180)	24-27 Years (%age out of 151)	>28 Years (%age out of 31)
	R (%age)	R (%age)	R (%age)
University Library	**138 (76.7%)**	**135 (89.4%)**	21 (67.7%)
Department	46 (25.6%)	49 (32.4%)	**22 (71%)**
Cybercafe	34 (18.9%)	40 (26.5%)	7 (22.6%)
Home	104 (57.8%)	60 (39.7%)	20 (64.5%)
Hostel	24 (13.3%)	42 (27.8%)	4 (12.9%)

Table 4.4 shows frequency distribution and percentage of data for male and female respondents for place of accessing DRI. Majority of male respondents (89.2%) accessed DRI from 'university library' whereas 76.6% female respondents accessed them from 'home'. The second preference for accessing DRI was found to be 'home' by 44.1% male respondents and it was 'university library' by 71.9% female respondents. It was found that majority of female respondents preferred to access DRI from their 'homes' as compared to male respondents who accessed them from 'university library' may be as the females have a tendency that they always feel more comfortable and secure at home.

Table 4.4: Place of Accessing Digital Resources of Information Gender-wise

Place of accessing digital resources of information	Male respondents (%age out of 195)	Female respondents (%age out of 167)
	R (% age)	R (% age)
University Library	**174 (89.2%)**	120 (71.9%)
Department	67 (34.4%)	69 (41.3%)
Cyber cafe	46 (23.6%)	35 (21%)
Home	86 (44.1%)	**128 (76.6%)**
Hostel	39 (20%)	34 (20.4%)

Table 4.5 shows frequency distribution and percentage of data of the three agricultural universities for place of accessing DRI. Majority of respondents in HAU (97.3%) and HPKV (81.1%) accessed DRI from 'university library' whereas 65.6% respondents from PAU accessed them from 'home'. The second preference for the same was found to be 'home' by the respondents of HAU and HPKV whereas it was 'university library' by PAU.

Table 4.5: Place of Accessing Digital Resources of Information University-wise

Place of accessing digital resources of information	HAU (% age out of 184)	HPKV (% age out of 53)	PAU (% age out of 125)
	R (%age)	R (%age)	R (%age)
University Library	**179 (97.3%)**	**43 (81.1%)**	72 (57.6%)
Department	54 (29.4%)	21 (39.6%)	42 (33.6%)
Cybercafe	46 (25%)	16 (30.2%)	19 (15.2%)
Home	77 (41.9%)	25 (47.2%)	**82 (65.6%)**
Hostel	39 (21.2%)	8 (15.1%)	23 (18.4%)

The Table 4.1 shows that majority of respondents from PAU access DRI 'daily' and Table 4.5 shows that they prefer to access it from their 'home'. Thus, it is clear that since Internet is all time accessible from homes that might be another reason that DRI were accessed daily by the majority of PAU respondents. Furthermore, it was found that the majority of respondents i.e. 158 (43.6%) used 'broadband' connection for internet at homes and/or in hostels to access DRI (Table 4.6).

Table 4.6: Type of Internet Connection at Home and/ or Hostel

Type of internet connection at home and/ or hostel	No. of respondents	Percentage
Broadband	**158**	**43.6%**
Leased line	3	0.8%
Wi-fi	30	8.3%
Dial up	16	4.4%
Cable	7	1.9%
ISDN (Integrated Services Digital Network)	0	0
DSL (Digital Subscribers Line)	3	0.8%
GPRS (General Packet Radio Service)	63	17.4%

Note: Percentage is out of 362

Similarly, the broadband was also reported by 76.2% respondents as the type of Internet connection used by their respective university. It was followed by 'leased line' as reported by 10.5% respondents (Table 4.7).

Table 4.7: Type of Internet Connection at University

Type of internet connection at university	No. of respondents	Percentage
Broadband	**276**	**76.2%**
Leased line	38	10.5%
Wi-fi	33	9.1%
Dial up	10	2.8%
Cable	7	1.9%
ISDN (Integrated Services Digital Network)	3	0.8%
DSL (Digital Subscribers Line)	9	2.5%
GPRS (General Packet Radio Service)	7	1.9%

Note: Percentage is out of 362

4.2.3 Adequacy of information accessed from digital resources of information

The response of 305 (84.2%) respondents in Fig. 4.3 shows that majority of respondents are competent enough to find 'adequate' information from DRI. A few of them, i.e. 52 (14.4%), are still there who would find 'inadequate' and very few i.e. 5 (1.4%) who would find 'very inadequate' information from DRI.

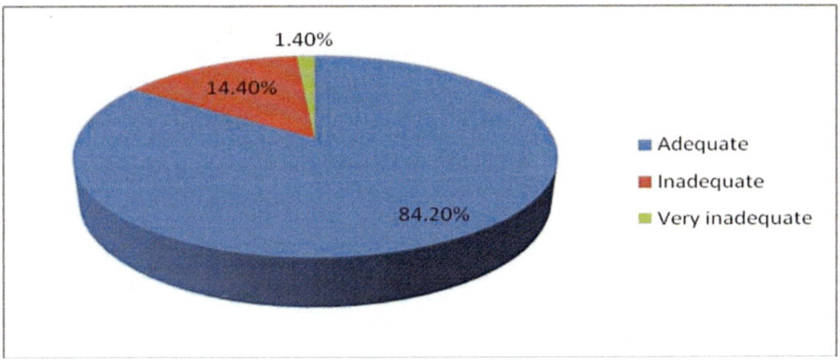

Note: Percentage is out of 362

Fig. 4.3: Adequacy of Information Accessed from Digital Resources of Information

Table 4.8 shows that majority of respondents i.e., 85.6%, 84.7% and 74.2% in the age groups 20-23 years, 24-27 years and more than 28 years respectively; adequately accessed information from DRI. However, a decrease in the quantum of information accessed as per their requirement was recorded as the age of respondents increased which shows the users from lower age groups are more competent in accessing information in required proportion from DRI.

Table 4.8: Adequacy of Information Accessed from Digital Resources of Information Age-wise

Adequacy of information accessed from digital resources of information	20-23 Years (% age out of 180)	24-27 Years (% age out of 151)	>28 Years (% age out of 31)
	R (%age)	R (%age)	R (%age)
Adequate	**154 (85.6%)**	**128 (84.7%)**	**23 (74.2%)**
Inadequate	25 (13.8%)	19 (12.6%)	8 (25.8%)
Very inadequate	1 (0.6%)	4 (2.7%)	0

Table 4.9 shows that 79.5% of male respondents and 89.8% of female respondents, accessed 'adequate' information from DRI. The percentage of female respondents was more than male respondents which shows the female respondents are more competent in accessing information as per required quantum from DRI.

Table 4.9: Adequacy of Information Accessed from Digital Resources of Information Gender-wise

Adequacy of information accessed from digital resources of information	Male respondents (%age out of 195)	Female respondents (%age out of 167)
	R (%age)	R (%age)
Adequate	**155 (79.5%)**	**[illegible] 8%)**
Inadequate	36 (18.4%)	16 (9.6%)
Very inadequate	4 (2.1%)	1 (0.6%)

Table 4.10 shows those 83.9% Master degree students and 85.3% PhD scholar's accessed 'adequate' information from DRI. The percentage of PhD scholars was more than Master degree students which shows the PhD scholars are more competent in accessing DRI in required proportions. The PhD scholars being more experienced and intensely involved in research may be more methodical in accessing DRI.

Table 4.10: Adequacy of Information Accessed from Digital Resources of Information Class-wise

Adequacy of information accessed from digital resources of information	Master degree students (%age out of 260)	PhD scholars (%age out of 102)
	R (%age)	R (%age)
Adequate	**218 (83.9%)**	**87 (85.3%)**
Inadequate	37 (14.2%)	15 (14.7%)
Very inadequate	5 (1.9%)	0

Table 4.8 revealed that percentage of respondents in the highest age group was less as compared to other two age groups in accessing the information from DRI adequately. Similarly, it was evident from the Table 4.10 that the percentage of Master degree students was comparatively less than PhD scholars in accessing the information from DRI adequately perhaps because Master degree students have comparatively less experience of using DRI than the PhD scholars. It is in accordance with Avdic and Eklund who focused on improving DIL skills amongst students of Orebro University of Sweden in order to enable them to use the various DRI subscribed by their library efficiently. For this purpose a survey was conducted on 150 students of the university. The problems perceived by students in using DRI were studied by an online survey based deductive approach with unified theory of acceptance and use of technology model as basis. The data

was quantitative and analyzed by calculating and comparing medians. It was found that students were mostly positive about their abilities but the results revealed that the information literacy (IL) could be much better. It was also found that older students experience more problems and so does less experienced ones (224-35). Furthermore, it is also evident from the Tables 4.8 to 4.10 that majority of respondents were able to access ‘ad[illegible]em accessed ‘inadequate’ and very few accessed ‘very ina[illegible]nformation from DRI. There is a need to enhance DIL skills as still there were few respondents who accessed ‘inadequate’ or ‘very inadequate’ information from DRI. This goal may be achieved possibly by conducting more user training programmes and extending need based assistance by library staff more extensively.

4.2.4 Mode of accessing digital resources of information

The respondents were enquired how they accessed the DRI subscribed by the library i.e. directly through the Uniform Resource Locator (URL) of DRI or through the website of their library. The Fig. 4.4 shows that 279 (77.1%) respondents that constitutes the major part, access DRI ‘through library website’.

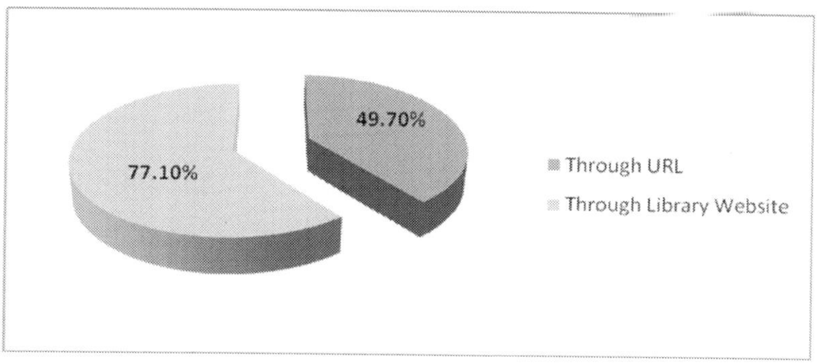

Note: Percentage is out of 362

Fig. 4.4: Accessing Digital Resources of Information Subscribed by the Library

In order to find out the DRI accessed by the respondents, three options were given to them as shown in Fig. 4.5. It was revealed that majority of the respondents i.e. 284 (78.4%) accessed ‘electronic databases subscribed by your university library’ followed by 251 (69.3%) who accessed ‘other digital resources available through Internet’. There were 178 (49.2%) respondents who accessed ‘electronic databases other than subscribed by your university library’. It is worth mentioning that the 'electronic databases other than subscribed by your university library' here refers to electronic databases to which a student gets access when he or she is a member of any library other than the library of the university he or she is pursuing studies. So, it was revealed that majority of students access the electronic databases subscribed by their University Library.

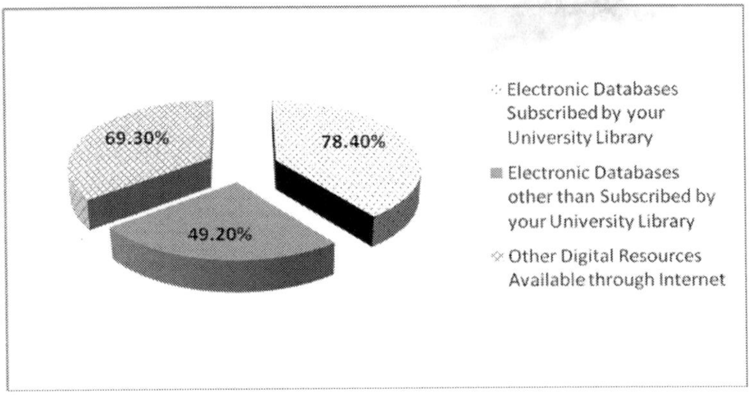

Note: Percentage is out of 362

Fig. 4.5: Accessing Digital Resources of Information

4.2.5 Mode of acquisition of information from digital resources of information

The ability of the respondents to access the DRI efficiently was also assessed by their capability of taking decision in complex situations. In this regard, an instance was given in which the respondents were provided with five options - 'pay the price of the article', 'try to find a person/library who has access to that article', 'try to contact the author and request for a free copy of the article', 'try to find the abstract of the article' and 'ignore it and search for some other freely available article'. They were asked to select the most appropriate option in case they come across a situation when an online article which is very important for their research work is available on payment basis. Fig. 4.6 shows that majority of respondents 128 (35.4%) would 'try to find the abstract of the article'. It was followed by 76 (21%) respondents who would 'try to find a person/library who has access to that article', 69 (19.1%) would 'ignore it and search for some other freely available article', 64 (17.7%) respondents would 'try to contact the author and request for a free copy of the article' and 25 (6.9%) respondents would 'pay the price of the article'.

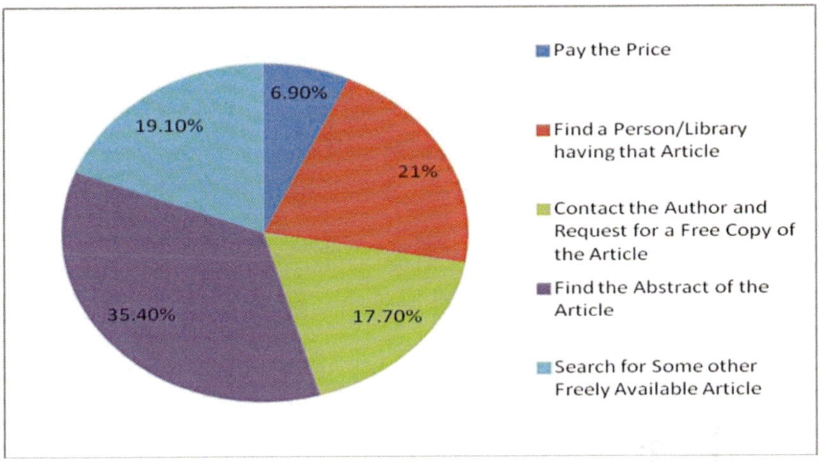

Note: Percentage is out of 362

Fig. 4.6: Accessing an Online Article Available on Payment

Table 4.11: Accessing an Online Article Available on Payment Age-wise

Accessing an online article available on payment	20-23 Years (%age out of 180)	24-27 Years (%age out of 151)	>28 Years (%age out of 31)
	R (%age)	**R (%age)**	**R (%age)**
Pay the Price of the Article	16 (8.9%)	8 (5.3%)	1 (3.2%)
Try to Find a Person/ Library who has Access to that Article	32 (17.8%)	36 (23.8%)	8 (25.8%)
Try to Contact the Author and Request for a Free Copy of the Article	19 (10.6%)	35 (23.2%)	**10 (32.3%)**
Try to Find the Abstract of the Article	**76 (42.2%)**	**44 (29.1%)**	8 (25.8%)
Ignore it and Search for Some other Freely Available Article	37 (20.6%)	28 (18.5%)	(12.9%)

The age-wise analysis shows that the majority of respondents i.e. 42.2% in the age group of 20-23 years and 29.1% of 24-27 years age would 'try to find the abstract of the article'. On the other hand, it was revealed that the majority of respondents i.e. 32.2% in the age group of more than 28 years would 'try to contact the author and request for a free copy of the article' (Table 4.11). Perhaps, the awareness about the other possible options like contacting the author of the article comes with age.

4.3 Searching Information from Digital Resources of Information

Various aspects of the ability of the subjects in searching information from DRI like search methods, preferences, keyword search, efficiency in searching relevant information, ease of searching information from DRI and time taken in searching information from DRI; was studied and is discussed in this section.

Table 4.12: Options Used for Searching Relevant Information

Options used for searching relevant information	Rank									W.I.	Rank as in W.I.
	1	2	3	4	5	6	7	8	9		
Refine your search by sorting them 'by relevance'	84	85	76	21	30	19	23	18	6	6.72	1
Refine your search by sorting chronologically	42	76	61	66	34	30	36	14	3	6.18	2
Refine your search by sorting highly cited	33	39	70	58	50	50	36	20	6	5.66	3
Going for advanced search	52	33	39	57	41	43	58	32	7	5.44	4
Consult the first 10 or 20 references	40	51	26	33	70	33	72	30	7	5.28	5
Randomly consult few of them	23	31	28	59	42	100	33	36	10	4.96	6
Consult all references one by one	64	22	36	37	17	38	54	59	35	4.91	7
Search again using some other keyword	17	18	25	32	62	30	33	121	24	4.03	8
Give up the search	7	6	2	0	14	21	16	32	264	1.81	9

4.3.1 Search Methods and Preferences

The responses regarding options used for searching relevant information out of the enormous information available from DRI was evaluated using W.I. and is shown in Table 4.12. It was found that majority of respondents refined their search 'by relevance' as this option got first rank in W.I. followed by 'refine search chronologically' and 'by sorting highly cited'. On the contrary, 'going for advanced search' got fourth rank. It shows that the respondents rely more on computers to sort out information and are lacking in DIL skills of advanced search and framing of appropriate keywords in order to search the information efficiently. Fowler and Dupuis reported same in their study however, introducing IL tutorial made remarkable impact on advanced search skills, framing keywords, searching indexes and evaluating websites (287-90). In order to achieve this, Digital Information Literacy Office at the University of Texas at Austin created Texas Information Literacy Tutorial (TILT). It was integrated with the existing library instructions programme. This online tutorial was designed to teach progression of skills and practice of basic IL skills through problem based interactions. A significant development in the IL skills of the students was observed as an impact of this tutorial.

Table 4.13: Digital Resources of Information Preferred for Accessing Full-text Articles

Digital resources of information preferred	Rank					W.I.	Rank as in W.I.
	1	2	3	4	5		
Google	203	69	52	32	6	4.19	1
Google Scholar	72	117	65	69	39	3.31	2
CeRA	66	71	101	83	41	3.11	3
ScienceDirect.com	16	81	112	118	35	2.79	4
Altavista	5	24	32	60	241	1.59	5

The preference with regard to the DRI for accessing full-text articles was also evaluated using W.I. Table 4.13 shows that majority of respondents used Google for accessing full-text articles followed by Google Scholar, CeRA and ScienceDirect.com as these DRI got first, second, third and fourth rank respectively in W.I. Altavista was the least preferred DRI amongst them as it was got fifth rank. The academic websites CeRA and ScienceDirect.com were found less preferred as compared to the search engines Google and Google Scholar. The reason might be that these search engines are easily accessible worldwide from the hostels and even homes whereas CeRA and ScienceDirect.com being subscribed sites are campus wide accessible only i.e. from libraries and departments. It is also worth mentioning that Google Scholar inspite of being academic site providing access to scholarly literature is preferred less as compared to Google thus more awareness needs to be generated amongst the respondents about such DRI which are exclusively academic in nature.

4.3.2 Keyword Searching

The competency of the respondents with regard to framing appropriate keywords while searching information from DRI was also evaluated by means of an example. A topic was given in the questionnaire along with five possible keywords. The respondents were asked to choose the most appropriate keyword out of the given options which they will use to search information on that topic from the three CDROM databases subscribed by their libraries. It is worth mentioning here that all the three agricultural universities are subscribing to three CDROM databases namely AGRICOLA, AGRIS and CAB. These databases are programmed by the proprietor to accept words 'and', 'or' and 'not' as connectors to join various terms to form relevant keywords. The word 'in' and punctuation mark 'hyphen' are not acceptable in these databases to be used as connectors in the keywords. Thus, the option 'apple and nutrition' was actually the most appropriate keyword here to search information on the topic 'nutritional value of apple' from the CDROM databases. The keywords i.e. 'apple or nutrition' and 'apple not nutrition' though technically acceptable to database but were not most appropriate as they would give larger number of references including considerable number of irrelevant ones also which need to be refined again in order to get the relevant references only. The data was analyzed and is shown in Fig. 4.7. It was found that 156 (43.2%) respondents were able to frame the most appropriate keyword i.e. 'apple and nutrition', which is even less than 50% of respondents. It was followed by 133 (36.7%) respondents who regarded 'nutrition in apple' as the most appropriate keyword, 66 (18.2%) considered 'apple-nutrition', 4 (1.1%) considered 'apple or nutrition' and 3 (0.8%) of respondents regarded 'apple not nutrition' as the most appropriate keyword.

Table 4.14 present age-wise frequencies and percentage of the ability of the respondents to frame appropriate keyword. It was found that majority of respondents i.e. 45% in the age group 20-23 years considered 'nutrition in apple' as most appropriate keyword. 49.7% in the age group 24-27 years and 58.1% of more than 28 years age regarded 'apple and nutrition' as most appropriate keyword. It may be noted that 'apple and nutrition' is actually the most appropriate keyword here, thus respondents of higher age groups are more competent in framing appropriate keywords and this skill increases with age. The youngest age group may be having comparatively less practice of using DRI are lacking behind in this skill thus need more attention of the library staff in enhancing it.

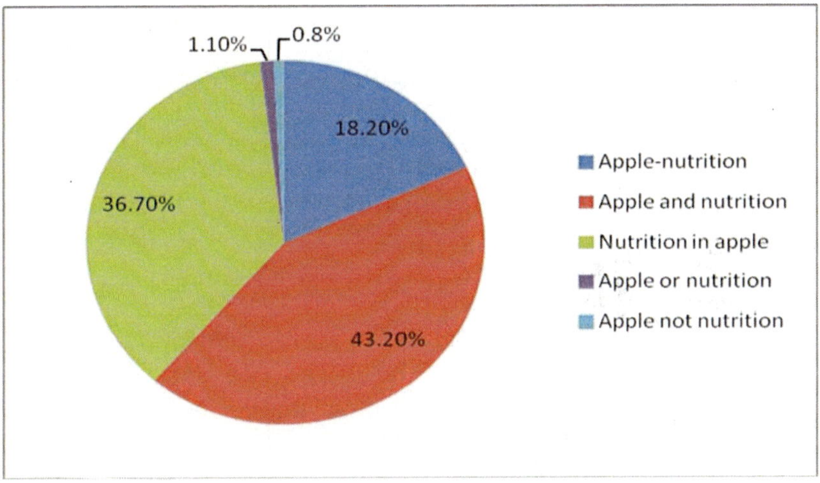

Note: Percentage is out of 362

Fig. 4.7: Keyword(s) Used for Searching CDROM Databases

Table 4.14: Keyword(s) Used for Searching CDROM Databases by Age

Keyword(s) used for searching CDROM database	20-23 Years (%age out of 180)	24-27 Years (%age out of 151)	>28 Years (%age out of 31)
	R (%age)	R (%age)	R (%age)
Apple-nutrition	33 (18.3%)	28 (18.5%)	5 (16.1%)
Apple and nutrition	63 (35%)	**75 (49.7%)**	**18 (58.1%)**
Nutrition in apple	**81 (45%)**	45 (29.8%)	7 (22.6%)
Apple or nutrition	3 (1.7%)	1 (0.7%)	0
Apple not nutrition	0	2 (1.3%)	1 (3.2%)

Table 4.15 shows frequency distribution and percentage of male and female respondents regarding their ability to frame appropriate keyword. It was found that majority of male respondents i.e. 51.8% considered 'apple and nutrition' as most appropriate keyword. 49.7%

female respondents regarded 'nutrition in apple' as most appropriate keyword. It shows that male respondents are more competent in framing appropriate keywords than female respondents.

Table 4.15: Keyword(s) Used for Searching CDROM Databases by Gender

Keyword(s) used for searching CDROM databases	**Male respondents (%age out of 195)**	**Female respondents (%age out of 167)**
	R (%age)	**R (%age)**
Apple-nutrition	39 (20%)	27 (16.2%)
Apple and nutrition	**101 (51.8%)**	55 (32.9%)
Nutrition in apple	50 (25.6%)	**83 (49.7%)**
Apple or nutrition	2 (1%)	2 (1.2%)
Apple not nutrition	3 (1.6%)	0

Table 4.16 gives class-wise frequencies and percentage of the ability of the respondents to frame appropriate keyword. It was found that 39.6% Master degree students regarded 'nutrition in apple' as most appropriate keyword followed by 38.8% who regarded 'apple and nutrition'. On the other hand, 53.9% PhD scholars considered 'apple and nutrition' as most appropriate keyword followed by 29.4% for 'nutrition in apple'. As already mentioned above that 'apple and nutrition' is actually the most appropriate keyword here, thus PhD scholars are more competent in framing appropriate keywords as compared to Master degree students may be due to having comparatively more number of years of using it.

Table 4.16: Keyword(s) Used for Searching CDROM Databases by Class

Keyword(s) used for searching CDROM databases	**Master degree students (%age out of 260)**	**PhD scholars (%age out of 102)**
	R (%age)	**R (%age)**
Apple-nutrition	51 (19.6%)	15 (14.7%)
Apple and nutrition	101 (38.8%)	**55 (53.9%)**
Nutrition in apple	**103 (39.6%)**	30 (29.4%)
Apple or nutrition	3 (1.2%)	1 (1%)
Apple not nutrition	2 (0.8%)	1 (1%)

Tables 4.12, 4.14 to 4.16 show the incompetency of the respondents in framing keywords. It is clear from Tables 4.14 to 4.16 that large numbers of respondents (about more than 50%) were not able to frame most appropriate keyword i.e. 'apple and nutrition' for using CDROM databases. This is in agreement with Brown and Krumholz who reported that students have difficulty in framing keywords, synonyms and related terms (111-23). However, an effort to overcome these weaknesses with regard to IL was made. Various IL skills of the students of University of Oklahoma Libraries like information seeking, evaluating and usage abilities were analyzed before, during and the end of a geo-microbiology course which included an IL component. 11% increase in IL skills of the respondents was recorded as an impact of this teacher librarian collaborative effort. Similarly, the incompetency in framing appropriate keywords reflected in the present study may also be overcome by imparting specific trainings to the users.

4.3.3 Efficiency in Searching Digital Resources of Information

The efficiency in searching relevant information from DRI was studied. The Fig. 4.8 shows that 279 (77.1%) respondents were able to find 'almost all' relevant information from DRI. 60 (16.6%) could find 'little', 20 (5.5%) were able to find 'all' relevant information from DRI. Only 3 (0.8%) were able to find 'very little' relevant information from DRI. By and large more than 82% of respondents were competent enough to find 'all' or 'almost all' relevant information, which is on higher side.

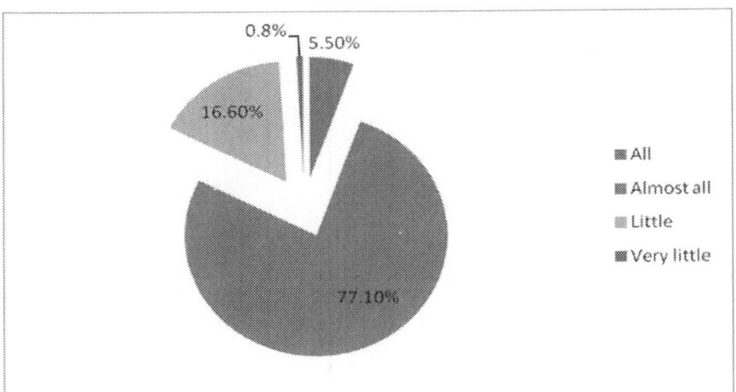

Note: Percentage is out of 362

Fig. 4.8: Efficiency in Searching Relevant Information from Digital Resources of Information

The efficiency in searching relevant information was also analyzed for different age groups and is presented in Table 4.17. It is clear that majority of respondents in all the three age groups - 78.9% in the age group of 20-23 years, 74.8% of 24-27 years and 77.4% of more than 28 years age, found 'almost all' the information accessed from DRI as relevant.

Table 4.17: Efficiency in Searching Relevant Information from Digital Resources of Information by Age

Efficiency in searching relevant information from digital resources of information	20-23 Years (%age out of 180)	24-27 Years (%age out of 151)	>28 Years (%age out of 31)
	R (%age)	R (%age)	R (%age)
All information	15 (8.3%)	5 (3.3%)	0
Almost all information	**142 (78.9%)**	**113 (74.8%)**	**24 (77.4%)**
Little information	23 (12.8%)	32 (21.2%)	5 (16.1%)
Very little information	0	1 (0.7%)	2 (6.5%)

The university and class-wise cross tabulation of the frequencies and percentage is given in Table 4.18. The majority of Master degree students - 67.6% from HAU, 77.8% from HPKV and 86.8% from PAU got 'almost all' relevant information from DRI. Similarly, majority of the PhD scholars - 88.9% from HAU, 76.5% from HPKV and 81.6% from PAU also got 'almost all' relevant information from DRI. It is also evident that incase of HAU, more number of PhD scholars were efficient in searching relevant information from DRI as compared to Master degree students, which shows that PhD scholars of HAU have better DIL skills in this case than Master degree students and it stands reverse in HPKV and PAU.

Though the Tables 4.17 and 4.18 show that the competency skills of about more than 76% of respondents in all the three age groups, universities and both the classes were satisfactory to find 'all' and 'almost all' relevant information from DRI but still there were considerable (around 23% maximum) respondents who were able to find 'little' and 'very little' relevant information which shows that DIL skills amongst the subjects are still in developing stage.

Table 4.18: University and Class-wise Tabulation of Efficiency in Searching Relevant Information from Digital Resources of Information

Efficiency in searching relevant information from digital resources of information	HAU		HPKV		PAU	
	Master degree students (%age out of 148)	PhD scholars (%age out of 36)	Master degree students (%age out of 36)	PhD scholars (%age out of 17)	Master degree students (%age out of 76)	PhD scholars (%age out of 49)
	R (%age)	R (%age)	R (%age)	R (%age)	R (%age)	R (%age)
All information	13 (8.8%)	0	2 (5.6%)	0	4 (5.3%)	1 (2%)
Almost all information	**100 (67.6%)**	**32 (88.9%)**	**28 (77.8%)**	**13 (76.5%)**	**66 (86.8%)**	**40 (81.6%)**
Little information	35 (23.6%)	4 (11.1%)	5 (13.9%)	4 (23.5%)	6 (7.9%)	6 (12.2%)
Very little information	0	0	1 (2.8%)	0	0	2 (4.1%)

4.3.4 Ease of Searching Information from Digital Resources of Information

The ease of searching information from DRI was evaluated and is presented in Fig. 4.9. There were 211 (58.3%) respondents for whom accessing DRI was 'easy' and it was 'very easy' for 68 (18.8%). However, 80 (22.1%) and 3 (0.8%) respondents were there for whom it was 'somewhat easy' and 'not easy' respectively.

The competence of searching information from DRI easily was evaluated age-wise and results are presented in Table 4.19. It was revealed that 49.4%, 69.6% and 54.8% in the age groups of 20-23 years, 24-27 years and above 28 years respectively searched information from DRI easily whereas for 25.6% respondents in the age group of 20-23 years and 22.6% of more than 28 years of age; it was 'very easy'. 19.2% and 22.6% respondents respectively in the age groups 24-27 years and more than 28 years; it was 'somewhat easy'.

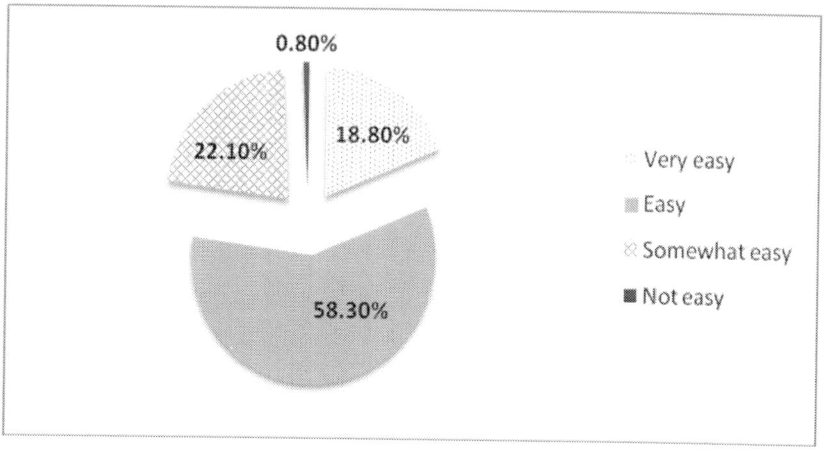

Note: Percentage is out of 362

Fig. 4.9: Ease of Searching Information from Digital Resources of Information

Table 4.19: Ease of Searching Information from Digital Resources of Information by Age

Ease of searching information from digital resources of information	20-23 Years (%age out of 180)	24-27 Years (%age out of 151)	>28 Years (%age out of 31)
	R (%age)	R (%age)	R (%age)
Very easy	46 (25.6%)	15 (9.9%)	7 (22.6%)
Easy	**89 (49.4%)**	**105 (69.6%)**	**17 (54.8%)**
Somewhat easy	44 (24.4%)	29 (19.2%)	7 (22.6%)
Not easy	1 (0.6%)	2 (1.3%)	0

The competence of searching information from DRI easily was evaluated and gender-wise results are presented in Table 4.20. It was found that 53.9% of male respondents and 63.5% of female respondents searched information from DRI easily whereas for 24.6% of male respondents and 19.1% of female respondents it was 'somewhat easy'. For 20.5% of male respondents and 16.8% of female respondents the same was 'very easy'. It was also evident that female respondents were able to access DRI more easily than male respondents thus exhibiting better DIL skills in this case.

Table 4.20: Ease of Searching Information from Digital Resources of Information by Gender

Ease of searching information from digital resources of information	Male respondents (%age out of 195)	Female respondents (%age out of 167)
	R (%age)	R (%age)
Very easy	40 (20.5%)	28 (16.8%)
Easy	**105 (53.9%)**	**106 (63.5%)**
Somewhat easy	48 (24.6%)	32 (19.1%)
Not easy	2 (1%)	1 (0.6%)

The ease of searching information from DRI was also analyzed class-wise and is presented in Table 4.21. It was found that for 53.1% of Master degree students and 71.6% of PhD scholars searching information was 'easy'. It was 'somewhat easy' for 24.6% Master degree students and 15.6% PhD scholars whereas for 21.1% Master degree students and 12.8% of PhD scholars it was 'very easy'. It was revealed that more PhD scholars were able to access it easily as compared to Master degree students, perhaps may be due to their having comparatively more years of practice into it. This shows that PhD scholars have better DIL skills in this case than Master degree students.

Table 4.21: Ease of Searching Information from Digital Resources of Information by Class

Ease of searching information from digital resources of information	Master degree students (%age out of 260)	PhD scholars (%age out of 102)
	R (%age)	R (%age)
Very easy	55 (21.1%)	13 (12.8%)
Easy	**138 (53.1%)**	**73 (71.6%)**
Somewhat easy	64 (24.6%)	16 (15.6%)
Not easy	3 (1.2%)	0

It is clear from Fig. 4.9, Tables 4.19 to 4.21 that searching information from DRI was 'easy' for majority of respondents of all the age groups as well as both the genders and classes and none was there who found it 'difficult'. But still there were around more than 20% respondents for whom searching information from DRI was 'somewhat easy' and 'not easy' which indicates that their proficiency needs to be improved probably by means of basic as well as specific need based training programmes along with special assistance provided by the library staff.

4.3.5 Time Taken in Searching Information from Digital Resources of Information

Number of days usually taken to search required information was also considered as a part of assessing the efficiency of the respondents in searching information from DRI. Fig 4.10 shows that 187 (51.7%) respondents were able to find required information from DRI within '2-5 days' followed by 81 (22.4%) in just 'one day', 72 (19.9%) in 'one week', 14 (3.8%) in '8-10 days' and 8 (2.2%) in 'more than 10 days'.

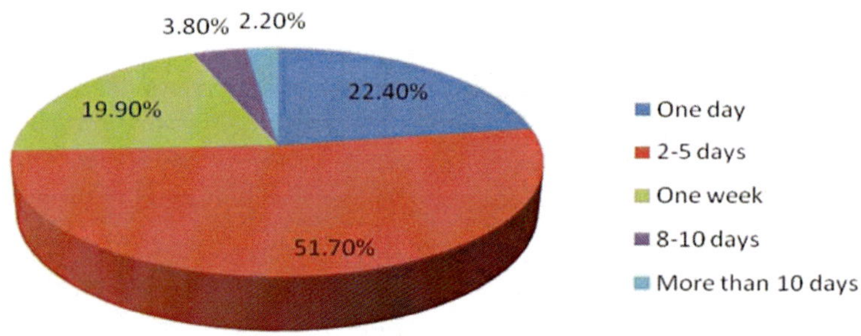

Note: Percentage is out of 362

Fig. 4.10: Total Number of Days Required to Search Information from Digital Resources of Information

The data was analyzed according to age of the respondents and is presented in the Table 4.22. The majority of respondents i.e. 50.6% in the age group of 20-23 years, 55% of 24-27 years age and 41.9% of more than 28 years age took '2-5 days' to search information from DRI. This was followed by 26.6% and 35.5% in the group of 20-23 years and above 28 years respectively who took only 'one day' whereas there were 23.2% in the age group of 24-27 years who took 'one week' for it.

Table 4.23 shows gender-wise response for the same. The majority of respondents- 49.2% male respondents and 54.5% female respondents took '2-5 days' for this purpose followed by 23.6% male respondents who took 'one week' as compared to 22.8% female respondents who took 'one day' only. In this instance, female respondents were found to be more efficient in searching information from DRI.

Table 4.22: Total Number of Days Required to Search Information from Digital Resources of Information by Age

Total number of days required to search information from digital resources of information	20-23 Years (%age out of 180)	24-27 Years (%age out of 151)	>28 Years (%age out of 31)
	R (%age)	R (%age)	R (%age)
One day	48 (26.6%)	22 (14.6%)	11 (35.5%)
2-5 days	**91 (50.6%)**	**83 (55%)**	**13 (41.9%)**
One week	33 (18.3%)	35 (23.2%)	4 (12.9%)
8-10 days	7 (3.9%)	4 (2.6%)	3 (9.7%)
More than 10 days	1 (0.6%)	7 (4.6%)	0

Table 4.23: Total Number of Days Required to Search Information from Digital Resources of Information by Gender

Total number of days required to search information from digital resources of information	Male respondents (%age out of 195)	Female respondents (%age out of 167)
	R (%age)	R (%age)
One day	43 (22.1%)	38 (22.8%)
2-5 days	**96 (49.2%)**	**91 (54.5%)**
One week	46 (23.6%)	26 (15.5%)
8-10 days	8 (4.1%)	6 (3.6%)
More than 10 days	2 (1%)	6 (3.6%)

The time usually taken to search information from DRI was also calculated class-wise and is presented in Table 4.24. It was observed that 50.4% Master degree students and 54.9% PhD scholars took '2-5 days' in order to search required information from DRI followed by 23.1% Master degree students took 'one week' for it whereas 25.5% of PhD scholars took only 'one day' for the same. In this case, PhD scholars are more capable in searching information from DRI again possibly due to their comparatively more experience of using the DRI.

Table 4.24: Total Number of Days Required to Search Information from Digital Resources of Information by Class

Total number of days required to search information from digital resources of information	Master degree students (%age out of 260)	PhD scholars (%age out of 102)
	R (%age)	R (%age)
One day	55 (21.2%)	26 (25.5%)
2-5 days	**131 (50.4%)**	**56 (54.9%)**
One week	60 (23.1%)	12 (11.8%)
8-10 days	10 (3.8%)	4 (3.9%)
More than 10 days	4 (1.5%)	4 (3.9%)

The Fig. 4.10, Tables 4.22 to 4.24 show that though the majority of respondents are competent enough to search required information from DRI in a short span of time i.e., five days or even less than that but still there were around 25% of respondents who took 'one week' or even more than that to search required information. This trend indicates that much more needs to be done in order to develop their DIL skills which may include assistance from library staff in using DRI, conducting more training programmes, library based assignments and references from DRI recommended by the faculty.

4.4 Selecting Authentic Digital Resources of Information

The proficiency of the subjects to select an authentic DRI was assessed through three different instances which are discussed in this section. Firstly, the respondents were asked to select the most authentic website out of the four options in order to search information on state-wise sex ratio in India. The Fig. 4.11 gives the response in general and shows that 244 (67.4%) respondents would use 'www.censusindia.gov.in' followed by 94 (26%) who would use 'www.india.gov.in', 12 (3.3%) who preferred 'www.cyberjournalist.org.in' and 'www.infochangeindia.org' each. It is worth mentioning here that the websites 'www.censusindia.gov.in' and 'www.india.gov.in' are authentic for accessing information on state-wise sex ratio in India. It shows that majority of respondents were competent in selecting an authentic DRI.

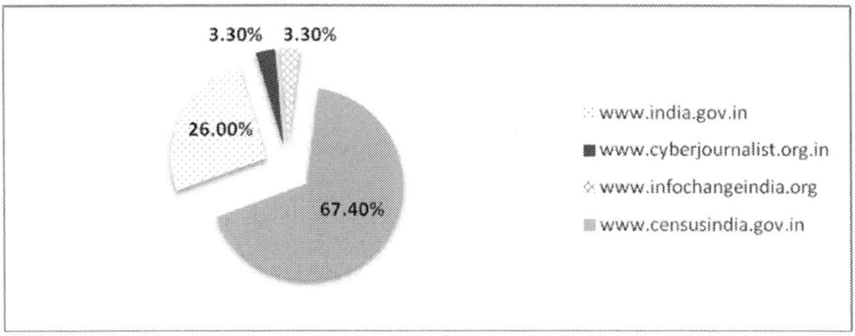

Note: Percentage is out of 362

Fig. 4.11: Most Authentic Digital Resource of Information

Table 4.25 shows age-wise frequency distribution and percentage to study the capability of respondents of evaluating DRI. It was observed that majority of respondents i.e., 73.3%, 59.6% and 71% respectively in the age group of 20-23 years, 24-27 years and above 28 years, regarded 'www.censusindia.gov.in' as most authentic website. It was followed by 'www.india.gov.in' which was regarded as most authenticate website by 21.1%, 31.8% and 25.8% respondents from the age group 20-23 years, 24-27 years and above 28 years respectively.

Table 4.25: Most Authentic Digital Resource of Information by Age

Most authentic digital resource of information	20-23 Years (%age out of 180)	24-27 Years (%age out of 151)	>28 Years (%age out of 31)
	R (%age)	R (%age)	R (%age)
www.india.gov.in	38 (21.1%)	48 (31.8%)	8 (25.8%)
www.cyberjournalist.org.in	5 (2.8%)	6 (4%)	1 (3.2%)
www.infochangeindia.org	5 (2.8%)	7 (4.6%)	0
www.censusindia.gov.in	**132 (73.3%)**	**90 (59.6%)**	**22 (71%)**

Table 4.26 shows results of the data analyzed gender-wise to study the ability of the respondents to evaluate a DRI with regard to its authenticity. It was found that the majority of male respondents (61.5%) and 74.3% of female respondents regarded 'www.censusindia.gov.in' as most authentic site followed by the website 'www.india.gov.in'. It was also observed that female respondents are more competent to evaluate a given DRI for its authenticity.

Table 4.26: Most Authentic Digital Resource of Information by Gender

Most authentic digital resource of information	Male respondents (%age out of 195)	Female respondents (%age out of 167)
	R (%age)	R (%age)
www.india.gov.in	65 (33.3%)	29 (17.4%)
www.cyberjournalist.org.in	4 (2.1%)	8 (4.8%)
www.infochangeindia.org	6 (3.1%)	6 (3.6%)
www.censusindia.gov.in	**120 (61.5%)**	**124 (74.3%)**

Table 4.27 shows that majority of respondents 63.6% from HAU, 66% from HPKV and 73.6% from PAU regarded 'www.censusindia.gov.in' as most authentic site followed by the website 'www.india.gov.in'. The respondents from PAU were found to be most competent in this case.

Table 4.27: Most Authentic Digital Resource of Information by University

Most authentic digital resource of information	HAU (%age out of 184)	HPKV (%age out of 53)	PAU (%age out of 125)
	R (%age)	R (%age)	R (%age)
www.india.gov.in	55 (29.9%)	12 (22.6%)	27 (21.6%)
www.cyberjournalist.org.in	6 (3.3%)	4 (7.6%)	2 (1.6%)
www.infochangeindia.org	6 (3.3%)	2 (3.8%)	4 (3.2%)
www.censusindia.gov.in	**117 (63.6%)**	**35 (66%)**	**92 (73.6%)**

Table 4.28 shows class-wise distribution of frequencies to study the ability of the respondents to evaluate a DRI with regard to its authenticity. It was found that majority of Master degree students (65.8%) and 71.6% of PhD scholars regarded 'www.censusindia.gov.in' as most authentic site followed by the website 'www.india.gov.in'. It was also revealed that PhD scholars are more competent in evaluating a DRI with regard to authenticity.

Table 4.28: Most Authentic Digital Resource of Information by Class

Most authentic digital resource of information	Master degree students (%age out of 260)	PhD scholars (%age out of 102)
	R (%age)	**R (%age)**
www.india.gov.in	73 (28.1%)	21 (20.6%)
www.cyberjournalist.org.in	7 (2.7%)	5 (4.9%)
www.infochangeindia.org	9 (3.5%)	3 (2.9%)
www.censusindia.gov.in	**171 (65.8%)**	**73 (71.6%)**

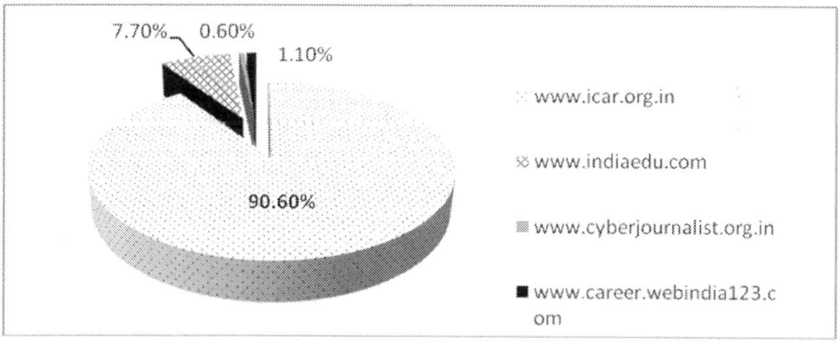

Note: Percentage is out of 362
Fig. 4.12: Best Digital Resource of Information

In order to assess the proficiency of the subjects to select an authentic DRI, another instance was used. The respondents were asked to select the best DRI out of given four options which they would use to find out total number of agricultural universities and research institutes in India. This question was followed by ranking of the reasons for selecting a particular DRI as best option by the respondents, which has been discussed later on in this section and is presented in Table 4.33. The analysis of data revealed that 'www.icar.org.in' was regarded as best DRI (Fig. 4.12). There were 328 (90.6%) respondents who regarded 'www.icar.org.in' as the best website for reasons discussed in Table 4.33 whereas, 28 (7.7%) respondents regarded 'www.indiaedu.com', only 4 (1.1%) held 'www.career.webindia123.com' and 2 (0.6%) marked 'www.cyberjournalist.org.in' as the best DRI. The website 'www.icar.org.in' is the official website of the Indian Council of Agricultural Research (ICAR), which is an apex body governing the Indian agricultural education and research system. All the agricultural universities and research institutes are established to function under

ICAR. Therefore, the website 'www.icar.org.in' is the best source to find out total number of agricultural universities and research institutes in India.

Table 4.29 shows that majority of respondents in all the three age groups i.e. 87.2%, 94.7% and 90.3%; regarded 'www.icar.org.in' as the best DRI.

Table 4.29: Best Digital Resource of Information by Age

Best digital resource of information	20-23 Years (%age out of 180)	24-27 Years (%age out of 151)	>28 Years (%age out of 31)
	R (%age)	R (%age)	R (%age)
www.icar.org.in	**157 (87.2%)**	**143 (94.7%)**	**28 (90.3%)**
www.indiaedu.com	18 (10%)	7 (4.6%)	3 (9.7%)
www.cyberjournalist.org.in	1 (0.6%)	1 (0.7%)	0
www.career.webindia123.com	4 (2.2%)	0	0

The website 'www.icar.org.in' was considered as the best DRI by the majority of male respondents (94.9%) and 85.6% of female respondents, as shown in Table 4.30. This shows that male respondents are more competent than female respondents in this regard.

Table 4.30: Best Digital Resource of Information by Gender

Best digital resource of information	Male respondents (%age out of 195)	Female respondents (%age out of 167)
	R (%age)	R (%age)
www.icar.org.in	**185 (94.9%)**	**143 (85.6%)**
www.indiaedu.com	9 (4.6%)	19 (11.4%)
www.cyberjournalist.org.in	0	2 (1.2%)
www.career.webindia123.com	1 (0.5%)	3 (1.8%)

Table 4.31 shows that 98.4% respondents from HAU, 83% from HPKV and 82.4% from PAU, which constituted the majority of respondents, have marked 'www.icar.org.in' as the best DRI. The students from HAU were found to be most competent in this case.

Table 4.31: Best Digital Resource of Information by University

Best digital resource of information	**HAU (%age out of 184)**	**HPKV (%age out of 53)**	**PAU (%age out of 125)**
	R (%age)	**R (%age)**	**R (%age)**
www.icar.org.in	**181 (98.4%)**	**44 (83%)**	**103 (82.4%)**
www.indiaedu.com	3 (1.6%)	7 (13.2%)	18 (14.4%)
www.cyberjournalist.org.in	0	1 (1.9%)	1 (0.8%)
www.career.webindia123.com	0	1 (1.9%)	3 (2.4%)

Table 4.32 shows that 90% Master degree students and 92.2% PhD scholars, which constituted the majority of respondents, have marked 'www.icar.org.in' as the best DRI. The PhD scholars were found to be more competent than Master degree students in this case.

Table 4.32: Best Digital Resource of Information by Class

Best digital resource of information	**Master degree students (%age out of 260)**	**PhD scholars (%age out of 102)**
	R (%age)	**R (%age)**
www.icar.org.in	**234 (90%)**	**94 (92.2%)**
www.indiaedu.com	20 (7.7%)	8 (7.8%)
www.cyberjournalist.org.in	2 (0.8%)	0
www.career.webindia123.com	4 (1.5%)	0

The selection criteria adopted by the respondents in selecting 'www.icar.org.in' as best DRI was also evaluated. The respondents were provided with various generally possible reasons usually used for selecting a DRI and they were asked to rank those options. The data was analyzed using W.I. It was found that first rank was given to 'authentic' followed by 'accessible' and 'regularly updated' by the respondents as shown in Table 4.33. The other options like 'highly used', 'relevant' and 'user friendly' got fourth, fifth and sixth rank respectively. Thus, the results show that authenticity was the major reasons for selecting the website 'www.icar.org.in' as best DRI to find out number of agricultural universities and research institutes in India by the respondents.

Table 4.33: Reasons for Best Digital Resource of Information

Reasons	Rank						W.I.	Rank as in W.I.
	1	2	3	4	5	6		
Authentic	106	96	45	43	32	40	4.22	1
Accessible	74	75	74	61	51	27	3.94	2
Regularly updated	75	65	69	65	60	28	3.85	3
Highly used	45	61	77	70	86	23	3.56	4
Relevant	34	54	69	83	98	24	3.37	5
User friendly	28	11	28	41	33	221	2.06	6

The ability of the respondents to select an authentic DRI was assessed through one more instance. In order to access research based information out of the websites of government organizations, educational institutions, commercial organizations and voluntary research/ other such organizations; the respondents were asked to rank these websites according to their authenticity. The data was calculated using W.I. Table 4.34 shows that websites of educational institutions got first rank in authenticity. This was followed by the websites of government, commercial and voluntary research/other such organizations.

Table 4.34 : Authenticity of Type of Websites

Websites	Rank				W.I.	Rank as in W.I.
	1	2	3	4		
Educational Institutions Websites (*.edu)	138	123	76	25	3.03	1
Government Websites (*.gov.in)	101	105	87	69	2.66	2
Commercial Websites (*.com)	81	51	97	133	2.22	3
Voluntary research/ other such organisation Websites (*.org)	42	83	102	135	2.09	4

Figures 4.11 and 4.12 and Tables 4.25 to 4.34 show that respondents are competent enough to evaluate the authenticity of a DRI or a website which is not in agreement with Flaspohler as well as Fowler and Dupuis who found that it was difficult for respondents to evaluate authenticity of a website and need additional practice to critically evaluate the information source they use. Their study dates back to year

2003 and 2000 respectively and ability of the users to evaluate authenticity of a website might have improved over time.

4.5 Conclusion

In nut shell it was found that the frequency of use of DRI was reported by only 29.6% of respondents as 'daily', which is not adequate. However, majority of respondents accessed 'electronic databases subscribed their university library' 'through library website' and were also able to find 'adequate' information from DRI. In order to search relevant information from the vast information available from DRI, the subjects preferred to refine search by sorting them 'by relevance' instead of going for 'advanced search option' and 'search again using some other keyword'. Around 57% respondents were not able to frame the most appropriate keyword in order to search relevant information from the DRI, which is not satisfactory. Majority of respondents were capable of finding 'almost all' and very few were able to find 'all' relevant information from DRI. The inability to frame appropriate keywords was regarded as the major reason of finding irrelevant information from DRI. This was followed by helplessness to use 'advanced search' option. On the other hand, majority of respondents were capable of searching information from DRI 'easily'. It was observed that majority of respondents took five days or even less than that to search information from DRI, which is satisfactory. It was also revealed that the subjects preferred 'Google' the most for accessing full-text articles as compared to the specialized websites like Google scholar, CeRA and ScienceDirect.com. There were adequate respondents who were able to evaluate the authenticity of a website. The websites of educational institutions were considered by the majority, as most authentic for research based information.

REFERENCES

Avdic, Anders and Anders Eklund. "Searching Reference Databases: What Students Experience and What Teachers Believe that Students Experience." *Journal of Librarianship and Information Science* 42.4 (2010): 224-35. *Sage Journals Online*. Web. 28 May 2012.

Brown, Cecelia, and Lee R. Krumholz. "Integrating Information Literacy into the Science Curriculum." *College and Research Libraries* 63.3 (2002): 111-23. Print.

Flaspohler, Molly R. "Information Literacy Program Assessment: One Small College Takes Big Plunge." *Reference Services Review* 31.2 (2003): 129-40. *Emerald.* Web. 14 Oct. 2009.

Fowler, Clara S., and Elizabeth A Dupuis. "What Have We Done? TILT's Impact on Our Instruction Program." *Reference Services Review* 28.4 (2000): 343-48. *Emerald.* Web. 12 Oct. 2009.

ROLE OF LIBRARIANS AND TEACHERS

5

5.1 Introduction

This chapter interprets the data analyzed for the problems faced by the users in using digital resources of information (DRI), support received from librarians and teachers in using DRI and users' perception about the competency of the library staff in terms of their training with regard to DRI. As already mentioned in previous chapters, the present study assesses digital information literacy (DIL) skills of the students by analyzing the data using MS Excel 2007 and Minitab 16 for frequency distribution, percentage, weighted index (W.I.), correlation and association of attributes.

5.2 Problems Faced in Using Digital Resources of Information

The problems faced by the subjects or in other words actual reasons due to which they found irrelevant information from DRI were also studied by calculating the data using W.I. The Table 5.1 shows that inability of the respondents to frame appropriate keywords was one of the major reasons followed by inability to use the 'advanced search' option and thirdly 'the journals covered in the databases are not useful'. 'Unable to refine the search' was also one of the reasons though it got fifth rank. 'Library staff is not helpful in searching information from digital sources' got sixth rank. Thus, once again, it is clear that respondents were not able to 'frame keywords', use 'advanced search' option and 'refine the search' which is in agreement with Fowler and Dupuis who reported same in their study. Fowler and Dupuis also informed that with introducing an IL tutorial, significant development in the IL skills like advanced search skills, framing keywords, searching indexes and evaluating websites; of the students were observed (287-90).

Table 5.1: Reasons of Finding Irrelevant Information from Digital Resources of Information

Reasons	Rank								
	1	2	3	4	5	6	7	W.I	Rank as in W.I.
Unable to frame appropriate key words	94	77	46	30	56	25	34	4.76	1
Unable to use 'advanced search' option	67	66	75	51	46	45	12	4.65	2
Journals covered in the databases are not useful	45	59	57	82	74	35	10	4.38	3
Sufficient number of digital resources are not available	53	55	62	75	56	26	35	4.33	4
Unable to refine the search	36	65	76	51	71	42	21	4.26	5
Library staff is not helpful in searching information from digital sources	25	23	32	36	22	124	100	2.85	6
Proper training is not imparted in this regard	42	17	41	37	37	65	150	2.78	7

The infrastructural or hardware related problems faced by the subjects were also evaluated by calculating the data using W.I. Table 5.2 show that 'virus infected computers' was one of the major reasons followed by 'slow internet connectivity' and 'inadequate number of computers'. 'Unavailability of printer' was also one of the reasons though it got fifth rank. The universities need to overcome these hurdles and take utmost care of the computer hardware, availability of adequate DRI and speed of the Internet connectivity so as to provide appropriate infrastructure in order to enable the optimum utilization of DRI and in turn enhancement of DIL skills of the users.

Table 5.2: Problems Faced while Accessing Information from Digital Resources of Information

Problems	Rank						W.I.	Rank as in W.I.
	1	2	3	4	5	6		
Virus infected computers	136	62	63	46	38	17	4.44	1
Slow Internet connectivity	121	59	30	70	42	40	4.07	2
Inadequate no. of computers	48	75	88	65	54	32	3.73	3
Saving data on pen drives or CDs	26	78	66	74	74	44	3.38	4
Unavailability of printer	16	56	77	72	96	45	3.14	5
Inadequate power back-up	15	32	38	35	58	184	2.23	6

5.3 Support from Library and Information Science Professionals and Teachers

The data pertaining to role of Library and Information Science (LIS) professionals as well as the teachers in helping their students to use DRI was analyzed in this section. The overall response of the data related to the help provided by library staff as given in Fig. 5.1 shows that 214 (59.1%) respondents regarded library staff as 'helpful' and 107 (29.6%) considered them 'very helpful' in using DRI. There were 41 (11.3%) respondents who regarded the library staff as 'not helpful' maybe due to their unconcerned and no work attitude.

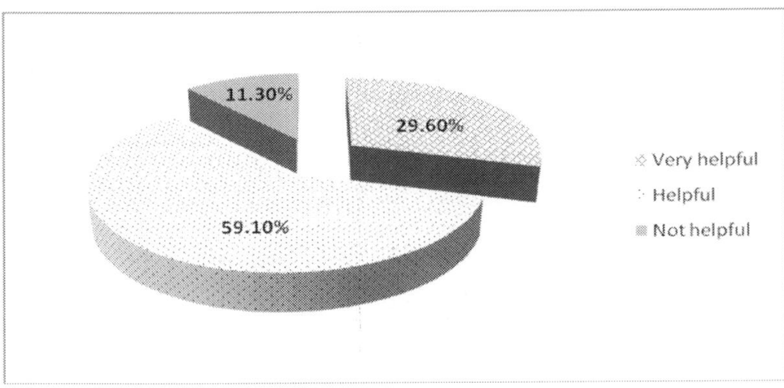

Note: Percentage is out of 362

Fig. 5.1: Help Provided by Library Staff in Using Digital Resources of Information

Table 5.3 shows data analyzed for male and female respondents. The library staff was found to be 'helpful' as indicated by 60% male respondents and 58.1% female respondents. 27.2% male respondents and 32.3% female respondents regarded library staff to be 'very helpful'. It was evident that library staff was regarded helpful by more male respondents than female respondents.

Table 5.3: Help Provided by Library Staff in Using Digital Resources of Information by Gender

Help provided by library staff in using digital resource of information	Male respondents (%age out of 195)	Female respondents (%age out of 167)
	R (%age)	R (%age)
Very helpful	53 (27.2%)	54 (32.3%)
Helpful	**117 (60%)**	**97 (58.1%)**
Not helpful	25(12.8%)	16 (9.6%)

Table 5.4 shows that library staff in all the three agricultural universities was found to be 'helpful' in using DRI. The same was indicated by 58.7% respondents of HAU, 52.8% of HPKV and 62.4% of PAU. The analysis of data revealed that library staff of PAU was most helpful and that of HPKV the least. 25% of respondents of HAU, 47.2% from HPKV and 28.8% from PAU had regarded library staff to be 'very helpful'. Only 16.3% from HAU and 8.8% from PAU held the same as 'not helpful'.

Table 5.4: Help Provided by Library Staff in Using Digital Resources of Information University-wise

Help provided by library staff in using digital resource of information	HAU (%age out of 184)	HPKV (%age out of 53)	PAU (%age out of 125)
	R (%age)	R (%age)	R (%age)
Very helpful	46 (25%)	25 (47.2%)	36 (28.8%)
Helpful	**108 (58.7%)**	**28 (52.8%)**	**78 (62.4%)**
Not helpful	30 (16.3%)	0	11 (8.8%)

All the three agricultural universities are teaching how to search information from DRI in the library and its services part of the technical writing course of Master degree students. In order to find the awareness of the respondents about this course they were asked whether their syllabus include something on searching information from DRI. The majority of respondents 278 (76.8%) responded 'yes' whereas 84 (23.2%) respondents responded 'no' to it (Table 5.5). This course is opted in any one semester of either Master or PhD degree. Perhaps, these 23.2% respondents were those who have still not opted for this course.

Table 5.5: Digital Resources of Information in Syllabus

Searching Information from Digital Resources of Information Included in Syllabus	No. of Respondents	Percentage
Yes	**278**	**76.8%**
No	84	23.2%

Note: Percentage is out of 362

The role of teachers in improving DIL skills of the students was determined by evaluating the number of digital references given by them to their students. The frequency distribution and percentage of the data in Table 5.6 shows that majority of the respondents, 131 (36.2%), reported that 21%-40% digital references were given by their teachers. It was followed by 107 (29.6%) respondents who reported 41%-60%, 82 (22.6%) reported 1%-20% and 39 (10.8%) respondents stated 61%-80% digital references were given by their teachers. Very few respondents, only 3 (0.8%), informed that 81%-100% digital references were given by their teachers. It is evident from the data that, more than 58% of the respondents reported that less than 40% of digital references were given by their teachers.

Table 5.6: Digital References Given by Teachers

Percentage of digital references given by teachers	No. of respondents	Percentage
1%-20%	82	22.6%
21%-40%	**131**	**36.2%**
41%-60%	107	29.6%
61%-80%	39	10.8%
81%-100%	3	0.8%

Note: Percentage is out of 362

The data was also analyzed university-wise and is presented in Table 5.7. It was found that majority 67 (36.4%) of respondents in

HAU, 22 (41.5%) from HPKV and 42 (33.6%) from PAU informed that 21%-40% digital references were given by their teachers. The percentage of digital references given by the teachers was found to be least in PAU and highest in case of HPKV.

Table 5.7: Digital References Given by Teachers University-wise

Percentage of digital references given by teachers	HAU (%age out of 184)	HPKV (%age out of 53)	PAU (%age out of 125)
	R (%age)	R (%age)	R (%age)
1%-20%	38 (20.6%)	9 (17%)	35 (28%)
21%-40%	**67 (36.4%)**	**22 (41.5%)**	**42 (33.6%)**
41%-60%	65 (35.3%)	18 (34%)	24 (19.2%)
61%-80%	14 (7.6%)	3 (5.7%)	22 (17.6%)
81%-100%	0	1 (1.9%)	2 (1.6%)

The Tables 5.6 and 5.7 show that the digital references given by the teachers were less than 40% as stated by around 58% of the respondents. Thus, more exhaustive efforts may be made by the teachers to improve the DIL skills of the students by encouraging the use of DRI.

Table 5.8: Factors Helpful in Using Digital Resources of Information

Factors helpful in using digital resources of information	Rank						Rank as in W.I.
	1	2	3	4	5	W.I.	
Self learning from the Internet/manuals, etc.	132	84	67	53	26	3.67	1
Orientation from teachers	48	101	94	90	29	3.14	2
Asking the peer group/seniors/ colleagues, etc.	61	101	59	61	80	3.01	3
Orientation from library staff	86	35	67	87	87	2.85	4
Orientation from database companies/suppliers/aggregators, etc.	35	41	75	71	140	2.34	5

The respondents were also investigated regarding factors which were helpful in using DRI. This data was analyzed using W.I. and 'self learning from the Internet/manuals, etc.' was reflected as most helpful factor as it got first rank shown in Table 5.8. It was followed by 'orientation from teachers', 'asking the peer group/seniors/colleagues, etc.' and conversely, 'orientation from library staff' got fourth rank. 'Orientation from database companies/suppliers/ aggregators, etc.' got fifth rank. Thus, the orientation programmes conducted by the libraries are not considered as helpful as other factors like self learning, teachers, peer group etc.

The importance of various options which were helpful in using DRI was also explored. This data was analyzed using W.I. and 'references of DRI given by teachers' was revealed as most important factor as it got first rank shown in Table 5.9 and it was followed by 'own efforts'. On the contrary, 'information provided in user orientation programmes' got third rank, 'assistance from library staff whenever need arises' got fourth rank followed by 'compulsory course on searching information from DRI' and 'help and instructions for using digital resources of information by teachers'. 'Special training sessions organized by the library' got last rank. It implies that efforts made by library staff are not considered as important and helpful in searching information from DRI as made by teachers and at own level of users.

Table 5.9: Importance of Options in Searching Information from Digital Resources of Information

Importance of options in searching information from digital resource of information	Rank							W.I	Rank as in W.I.
	1	2	3	4	5	6	7		
References of DRI given by your teachers	55	72	73	53	62	31	16	4.58	1
Own efforts	125	17	27	42	17	24	110	4.11	2
Information provided in 'User Orientation Programmes'	31	60	40	86	56	65	24	3.99	3
Assistance from library staff whenever need arises	48	53	61	36	55	61	48	3.97	4
Compulsory course on searching information from DRI	53	30	65	54	49	45	66	3.85	5
Help and instructions for using DRI by teachers	22	72	60	36	53	68	51	3.81	6
Special training sessions organized by the library	28	58	36	55	70	68	47	3.69	7

Tables 5.8 and 5.9, shows that the training programmes conducted by the libraries were not adequate to cope up with the challenges of DIL. Thus the libraries need to reframe their user training programmes.

5.4 Training of Library and Information Science Professionals

The users' perception about the competency of the library staff was also evaluated. The total response of respondents about training of the library staff with regard to DRI is presented in the Fig. 5.2. The library staff was regarded as 'average trained' by 177 (48.9%) respondents and 172 (47.5%) considered them 'well trained'. Perhaps the library staff may have received trainings, attended various workshops and had discussions with peer group but still they need to update themselves constantly to render their services more efficiently to the users in order to improve their DIL skills. On the other hand, only 13 (3.6%) regarded the library staff as 'not trained'.

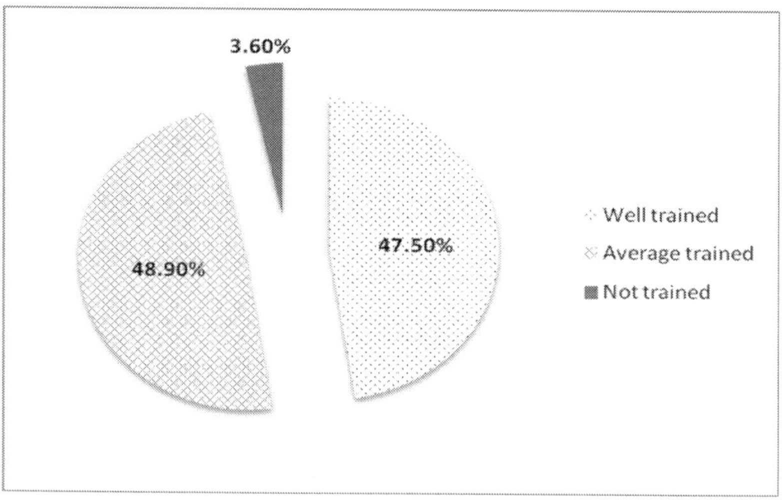

Note: Percentage is out of 362

Fig. 5.2: Training of Library Staff with regard to Digital Resources of Information

Table 5.10 shows the frequency distribution and percentage of respondents in all the three age groups. It was found that majority of respondents (52.8%) in the age group of 20-23 years regarded their library staff as 'well trained'. Whereas, the majority i.e., 51.7% and 61.3% in the age groups of 24-27 years and more than 28 years respectively, considered their library staff as 'average trained'. 44.4%

respondents of 20-23 years of age considered their library staff as 'average trained' on the other hand 45% respondents in the age group of 24-27 years and 29% of more than 28 years of age regarded them 'well trained'. Only 2.8%, 3.3% and 9.7% regarded them as 'not trained'. It is also clear that library staff was regarded 'well trained' by more respondents from lower age group probably they have received desirable assistance from library staff in using DRI or they have established better communication with them.

Table 5.10: Training of Library Staff with regard to Digital Resources of Information by Age

Training of library staff with regard to digital resource of information	20-23 Years (%age out of 180)	24-27 Years (%age out of 151)	>28 Years (%age out of 31)
	R (%age)	R (%age)	R (%age)
Well trained	**95 (52.8%)**	68 (45%)	9 (29%)
Average trained	80 (44.4%)	**78 (51.7%)**	**19 (61.3%)**
Not-trained	5 (2.8%)	5 (3.3%)	3 (9.7%)

The same information was also explored for all the three universities. Table 5.11 shows the frequency distribution and percentage of respondents. It was found that majority of respondents of HAU i.e. 54.4% and HPKV (51%) considered their library staff as 'well trained'. In contrast, 57.6% from PAU considered them 'average trained' maybe the library staff was not given suitable training in order to equip themselves in helping the users in using DRI as well as enhancement of DIL skills of the users. The attitude of no work and heavy workload due to high user staff ratio may be other reasons. On the other hand, only 2.7% from HAU and 6.4% from PAU considered them 'not trained'; which is a very low percentage.

It is clear from Fig. 5.2, Tables 5.10 and 5.11 that around 45% respondents regarded the library staff as 'average trained'. Thus, it can be concluded that more trainings regarding the use of DRI needs to be provided to the library staff so that they are capable of enhancing DIL skills of the users which supports Appleton who also advocates that more trainings regarding use of DRI need to be imparted to the library staff so that they may improve their skills and comprehension about DIL. Such trainings would enable them to support the DIL needs of students in a better way. Appleton, discussed about an online IL training programme called LolliPop used as a staff development tool in West

Cheshire College, England. An increase in confidence, competence and understanding of the DIL skills was observed amongst the staff members who had undergone this training. Therefore, they were better equipped to support the DIL requirements of users (191-98).

Table 5.11: Training of Library Staff with regard to Digital Resources of Information by University

Training of library staff with regard to digital resource of information	HAU (%age out of 184)	HPKV (%age out of 53)	PAU (%age out of 125)
	R (%age)	R (%age)	R (%age)
Well trained	**100 (54.4%)**	**27 (51%)**	45 (36%)
Average trained	79 (42.9%)	26 (49.1%)	**72 (57.6%)**
Not-trained	5 (2.7%)	0	8 (6.4%)

5.4 Conclusion

In nut shell, the major infrastructural or hardware related problems faced by the users while accessing DRI was 'Virus infected computers' followed by 'slow internet connectivity'. As far as support from LIS professionals is concerned the library staff was regarded 'very helpful' by around one-third of respondents whereas there were sufficient respondents who considered them 'helpful'. The library staff was regarded as 'very helpful' by maximum respondents from HPKV and the least by HAU. The library staff was perceived as 'average trained' by majority respondents. 'Self learning from internet/manuals, etc.' was regarded as most helpful factor in using DRI as against 'orientation from library staff'. 'References of digital resources of information given by teachers' were considered as most important factor in referring information from DRI as compared to various efforts made by the library staff. However, it was revealed that more than 58% of the respondents reported that less than 50% of digital references were given by their teachers. The percentage of digital references given by the teachers was found to be least in PAU and highest in case of HPKV.

REFERENCES

Appleton, Leo. "LolliPop for Learning Resources: Information Literacy Staff Training within Further Education." *Journal of Librarianship and Information Science* 42.3 (2010): 191-98. *Sage Journals Online*. Web. 29 May 2012.

Fowler, Clara S., and Elizabeth A Dupuis. "What Have We Done? TILT's Impact on Our Instruction Program." *Reference Services Review* 28.4 (2000): 343-48. *Emerald.* Web. 12 Oct. 2009.

■■■■

DIGITAL INFORMATION LITERACY IN SOCIAL LIVES

6

6.1 Introduction

Every person who possesses digital information literacy (DIL) skills understands the economic, legal, and social issues of using digital resources of information (DRI) and accesses and uses them accordingly. DIL skills form the basis for lifelong learning and extend learning beyond formal education. The individuals possessing DIL skills are capable of applying these skills in their jobs as well as in their social lives. The present chapter analyses the data and interprets it for adoption of DIL skills by the students in their day to day social life.

6.2 Digital Information Literacy in Social Lives

Digital information literacy in the social lives of the respondents i.e., use of information from DRI in daily routine life of the respondents was evaluated. Their DIL competency level was tested on five social situations such as railway reservations, bank transactions, job searching, format of newspaper read and immigration. The mode of railway reservation was taken as one case in this point and data was analyzed according to age and gender. First of all, the overall response of the data is presented in Fig 6.1. It shows that 268 (74%) of respondents would totally opt for DRI as they opted to 'check from website and reserve online'. 75 (21%) respondents would 'check from website and reserve from reservation counter'. Only 10 (2.8%) would go to 'reservation counter' and 9 (2.5%) respondents would 'request a friend'. This shows that majority of respondents use DRI in general which shows acceptance of DIL in their social lives.

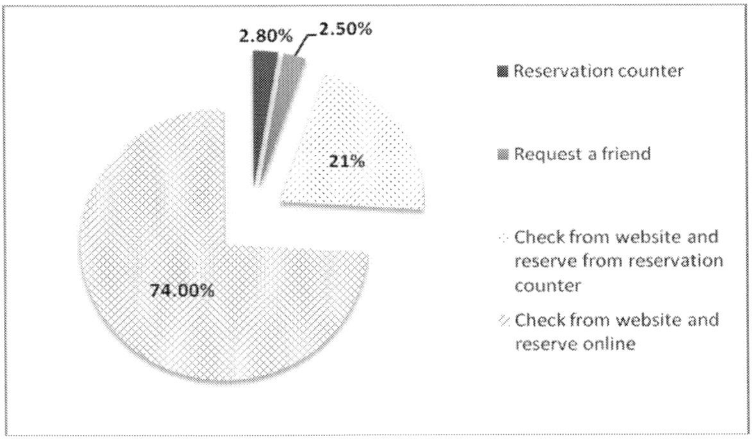

Note: Percentage is out of 362

Fig. 6.1: Mode of Railway Reservation to Attend a Conference

Table 6.1 shows age-wise frequency distribution of data. It was found that majority of students in all the age groups - 71.1% respondents in 20-23 years age, 76.8% in 24-27 years and 77.4% in more than 28 years age group; checked the status from website and reserved the ticket online. This tendency was found to increase with age perhaps the user from higher age groups had better understanding of the benefits of using DRI in their academic as well as social lives. There were few respondents - 25.6% in the age group of 20-23 years, 15.9% in 24-27 years and 16.1% of more than 28 years age who checked the reservation status from website and then go to reservation counter to reserve a ticket.

Table 6.1: Mode of Railway Reservation to Attend a Conference by Age

Mode of railway reservation to attend a conference	20-23 Years (%age out of 180)	24-27 Years (%age out of 151)	>28 Years (%age out of 31)
	R (%age)	R (%age)	R (%age)
Reservation counter	5 (2.8%)	4 (2.7%)	1 (3.2%)
Request a friend	1 (0.6%)	7 (4.6%)	1 (3.2%)
Check from website and reserve from reservation counter	46 (25.6%)	24 (15.9%)	5 (16.1%)
Check from website and reserve online	**128 (71.1%)**	**116 (76.8%)**	**24 (77.4%)**

Table 6.2 show gender-wise frequency distribution and percentage. Majority of male respondents i.e. 74.4% and 73.7% female respondents checked the status from website and reserved the ticket online followed by only 19.5% male respondents and 22.1% female respondents who checked the reservation status from website and then go to reservation counter to reserve a ticket. It also shows that there were more male respondents than female who used digital mode for railway reservation.

Table 6.2: Mode of Railway Reservation to Attend a Conference by Gender

Mode of railway reservation to attend a conference	Male respondents (%age out of 195)	Female respondents (%age out of 167)
	R (%age)	R (%age)
Reservation counter	5 (2.6%)	5 (3%)
Request a friend	7 (3.6%)	2 (1.2%)
Check from website and reserve from reservation counter	38 (19.5%)	37 (22.1%)
Check from website and reserve online	**145 (74.4%)**	**123 (73.7%)**

Digital information literacy in the social lives of male or female respondents was also assessed by their trend of using manual or digital means in withdrawal of money from bank. The Fig. 6.2 shows that 338 (93.4%) of respondents used digital means i.e. Automatic Teller Machine (ATM) in order to withdraw money from bank. This shows adoption of DIL in their social lives. There were only 13 (3.6%) respondents who opted for 'withdrawal form' and 11 (3%) preferred 'cheque' as a mode of withdrawal of money from bank.

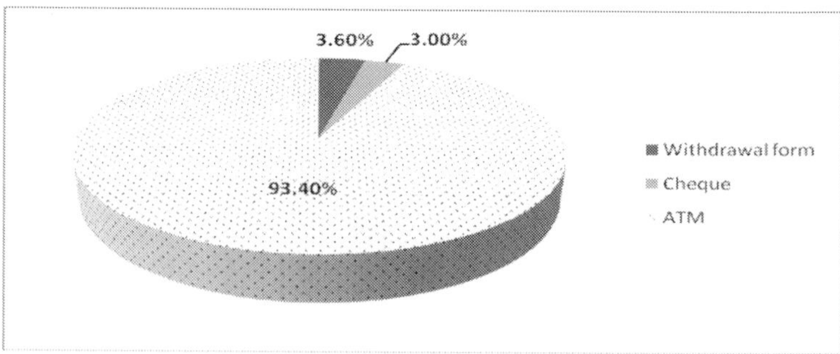

Note: Percentage is out of 362

Fig. 6.2: Mode of Withdrawal of Money to Deposit Semester Fees

Table 6.3 shows that 92.8% of respondents in the age group of 20-23 years, 93.4% in age group of 24-27 years age and 96.8% of more than 28 years of age used ATM for withdrawal of money from banks. In this case, it was revealed that the use of DIL in social lives increases with age. It is also evident from the table that very low percentage of respondents used other means like cheque and withdrawal form for the same.

Table 6.3: Mode of Withdrawal of Money to Deposit Semester Fees by Age

Mode of withdrawal of money	**20-23 Years (%age out of 180)**	**24-27 Years (%age out of 151)**	**>28 Years (%age out of 31)**
	R (%age)	**R (%age)**	**R (%age)**
Withdrawal form	6 (3.3%)	6 (3.9%)	1 (3.2%)
Cheque	7 (3.9%)	4 (2.7%)	0
ATM	**167 (92.8%)**	**141 (93.4%)**	**30 (96.8%)**

Table 6.4 shows gender-wise frequency and percentage of data. It was found that 93.3% male respondents and 93.4% female respondents used ATM to withdraw money from bank. However, very low percentage of both male and female respondents were found using other methods like cheque and withdrawal form in order to withdraw money from the bank. In this instance there was negligible difference (0.1%) between male respondents and female respondents in using ATM cards to withdraw money.

Table 6.4: Mode of Withdrawal of Money to Deposit Semester Fees by Gender

Mode of withdrawal of money	**Male respondents (%age out of 195)**	**Female respondents (%age out of 167)**
	R (%age)	**R (%age)**
Withdrawal form	8 (4.1%)	5 (3%)
Cheque	5 (2.6%)	6 (3.6%)
ATM	182 (93.3%)	156 (93.4%)

Another instance used to evaluate DIL in social lives was sources consulted by respondents for searching a job. The Fig. 6.3 shows that 'employment news' was most preferred source followed by 'Internet' for searching a job opportunity. 272 (75.1%) respondents preferred 'employment news' and 191 (53%) 'newspapers' as compared to 259 (72%) preferred 'internet', 224 (61.9%) would 'register online in various job sites' and 219 (60.5%) for 'visiting the websites of the concerned companies'. 'Placement agency' was least preferred option by 102 (28.2%) respondents.

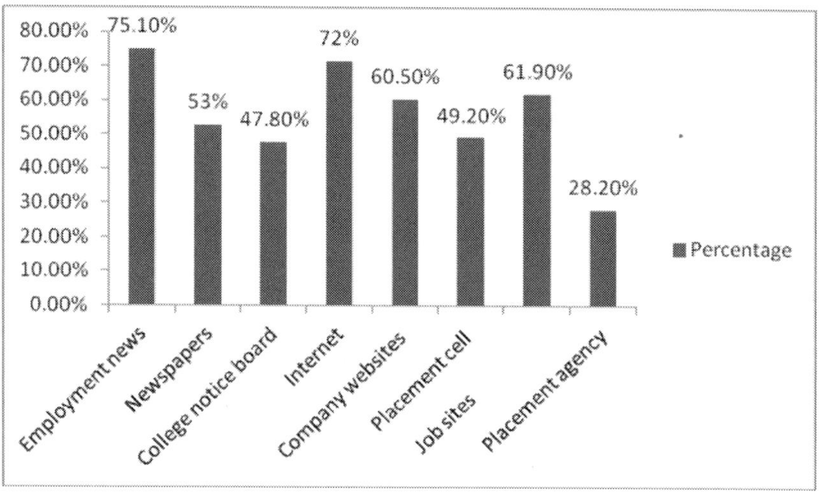

Note: Percentage is out of 362

Fig. 6.3: Sources Consulted for Searching a Job

Table 6.5 shows that 'Internet' was used by majority of respondents, 69.4%, in the age group of 20-23 years followed by 67.8% who used 'employment news' and 63.9% 'registered online in various job sites'. On the contrary, 82.1% respondents constituting a major part, in the age group 24-27 years used 'employment news' followed by 74.2% who used 'Internet' and 62.3%'visited the websites of the concerned companies'. Similarly, 83.9% respondents of more than 28 years of age also used 'employment news', 71% each used 'newspapers' and 'Internet', and 67.7% respondents each 'registered online in various job sites'. It is evident that older respondents prefer to use print media i.e., 'employment news' over the internet.

Table 6.5: Sources Consulted for Searching a Job by Age

Sources consulted for searching a job	20-23 Years (%age out of 180)	24-27 Years (%age out of 151)	>28 Years (%age out of 31)
	R (%age)	R (%age)	R (%age)
Employment news	122 (67.8%)	**124 (82.1%)**	**26 (83.9%)**
Newspapers	84 (46.7%)	85 (56.3%)	22 (71%)
College notice board	80 (44.4%)	78 (51.7%)	15 (48.4%)
Internet	**125 (69.4%)**	112 (74.2%)	22(71%)
Visiting the web-sites of the conce-rned companies	105 (58.3%)	94 (62.3%)	20 (64.5%)
Enquiring the placement cell of your institution	92 (51.1%)	70 (46.4%)	16 (51.6%)
Register online in various job sites	115 (63.9%)	88 (58.3%)	21(67.7%)
Placement agency	(26.7%)	(29.1%)	10 (32.3%)

Table 6.6 shows same trend in both male and female respondents. 78.5% male respondents used 'employment news'; 75.9% male respondents used 'Internet' and 61.5% each 'visited the websites of the concerned companies' and 'registered online in various job sites'. Similarly, 71.3% female respondents used 'employment news', 66.5% used 'Internet' and 62.3%'registered online in various job sites'.

Table 6.7 shows university-wise distribution of frequency regarding consultation of various sources in order to search a job. It was evident that majority (80.4%) respondents in HAU and 75.5% from HPKV used 'employment news' followed by 'Internet' for the above said purpose. On the other hand majority (69.6%) respondents of PAU used 'Internet' followed by 'employment news' for the same. In this instance, it was revealed that majority of respondents from PAU prefer DRI than print media as compared to majority of respondents from HAU and HPKV.

Table 6.6: Sources Consulted for Searching a Job by Gender

Sources consulted for searching a job	Male Respondents (% age out of 195)	Female Respondents (% age out of 167)
	R (%age)	R (%age)
Employment news	**153 (78.5%)**	**119 (71.3%)**
Newspapers	108 (55.4%)	83 (49.7%)
College notice board	96 (49.2%)	77 (46.1%)
Internet	148 (75.9%)	111 (66.5%)
Visiting the websites of the concerned companies	120 (61.5%)	99 (59.3%)
Enquiring the placement cell of your institution	105 (53.9%)	73 (43.7%)
Register online in various job sites	120 (61.5%)	104 (62.3%)
Placement agency	60 (30.8%)	42 (25.2%)

Table 6.7: Sources Consulted for Searching a Job by University

Sources consulted for searching a job	HAU (%age out of 184)	HPKV (%age out of 53)	PAU (%age out of 125)
	R (%age)	R (%age)	R (%age)
Employment news	**148 (80.4%)**	**40 (75.5%)**	84 (67.2%)
Newspapers	98 (53.3%)	22 (41.5%)	71 (56.8%)
College notice board	97 (52.7%)	24 (45.3%)	52 (41.6%)
Internet	141 (76.6%)	31 (58.5%)	**87 (69.6%)**
Visiting the websites of the concerned companies	114 (62%)	27 (50.9%)	78 (62.4%)
Enquiring the placement cell of your institution	102 (55.4%)	19 (35.9%)	57 (45.6%)
Register online in various job sites	123 (66.9%)	30 (56.6%)	71 (56.8%)
Placement agency	61 (33.2%)	17 (32.1%)	24 (19.2%)

Table 6.8 shows that majority of Master degree students (70.8%) as well as PhD scholars (86.3%) used 'employment news' for searching a job followed by 'Internet'.

Table 6.8: Sources Consulted for Searching a Job by Class

Sources consulted for searching a job	Master degree students (%age out of 260)	PhD scholars (%age out of 102)
	R (%age)	R (%age)
Employment news	**184 (70.8%)**	**88 (86.3%)**
Newspapers	124 (47.7%)	67 (65.7%)
College notice board	114 (43.9%)	59 (57.8%)
Internet	179 (68.9%)	80 (78.4%)
Visiting the websites of the concerned companies	148 (56.9%)	71 (69.6%)
Enquiring the placement cell of your institution	127 (48.9%)	51 (50%)
Register online in various job sites	160 (61.5%)	64 (62.8%)
Placement agency	70 (26.9%)	32 (31.4%)

The preferences of the respondents regarding the format in which they read newspapers were also ascertained. Fig. 6.4 shows that 222 (61.3%) respondents which form a major part prefer to read newspapers in 'print' format. This was followed by 113 (31.2%) respondents who prefer to read newspapers in 'both' the formats i.e. print as well as online. Only 27 (7.5%) respondents prefer to read 'online' format. This trend shows that newspapers are still preferred in print format as compared to online.

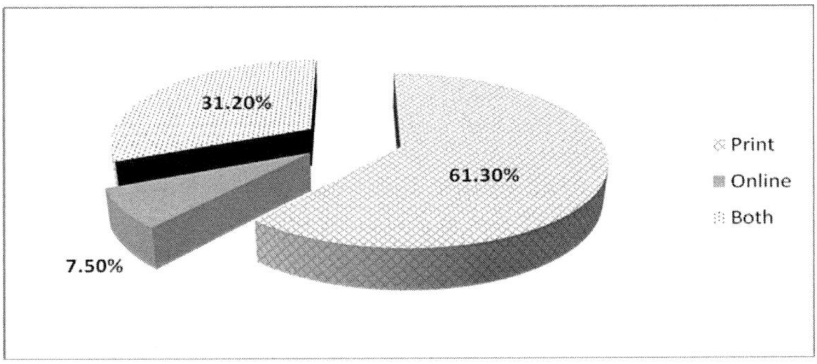

Note: Percentage is out of 362

Fig. 6.4: Format of Newspapers

The university-wise trend shows that majority of respondents 119 (64.7%) in HAU and 84 (67.2%) in PAU prefer to read newspapers in 'print' format. However, it was 'both' print and online format preferred by the majority of respondents i.e. 32 (60.4%) incase of HPKV (Table 6.9). HPKV is situated in hilly area and sometimes newspapers reach late in hilly areas during rainy and winter seasons. Perhaps, the respondents there may read newspaper in 'online' format during such conditions.

Table 6.9: Format of Newspapers University-wise

Format of newspapers	HAU (%age out of 184)	HPKV (%age out of 53)	PAU (%age out of 125)
	R (%age)	R (%age)	R (%age)
Print	**119 (64.7%)**	19 (35.8%)	**84 (67.2%)**
Online	20 (10.9%)	2 (3.8%)	5 (4%)
Both	45 (24.5%)	**32 (60.4%)**	36 (28.8%)

The adoption of DIL by the respondents in their social lives was also ascertained by the instance of the sources consulted by them for going abroad. Fig. 6.5 shows that majority of respondents 303 (83.7%) would 'search for relevant information from the specific website'. 190 (52.5%) respondents would 'consult teachers', 147 (40.6%) respondents would 'go to immigration consultancy' and 144 (39.8%) would 'consult friends'. In this instance also, it is clear that the respondents have well adopted DIL in their social lives.

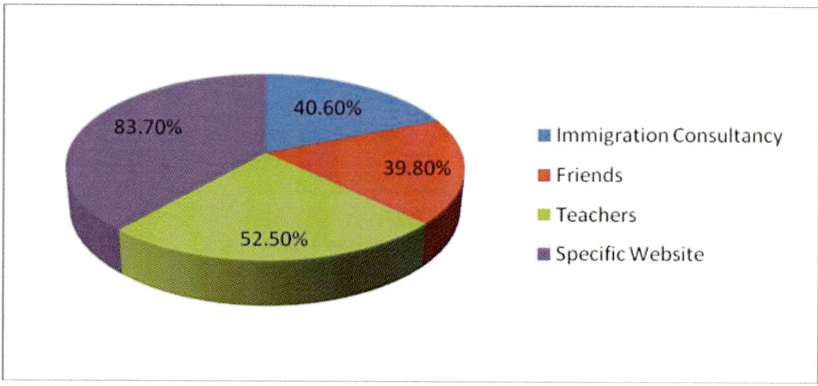

Note: Percentage is out of 362

Fig. 6.5: Sources Consulted for Going Abroad

The data presented in the Fig. 6.1, 6.2, 6.5 and Tables 6.1 to 6.4 show that the majority of respondents except a very few were efficiently using DIL in their social lives. However, Fig. 6.3, 6.4 and Tables 6.5 to 6.9 show that 'employment news' and newspapers in 'print' format are still preferred over digital media. Perhaps, 'employment news' is published since 1976 and is considered as more reliable source for employment opportunities. It provides employment related information at a single platform. On the other hand, Internet though provides access to enormous job related sites but reliability of many such sites is doubtful. Moreover, information on internet needs to be filtered first.

6.3 Suggestions for Improving Efficiency in Using Digital Resources of Information

Various suggestions to improve the efficiency in using DRI amongst the subjects were also studied. The Fig. 6.6 shows that overall 260 (71.8%) respondents suggested that 'libraries should more frequently organize training programmes in this regard' followed by 251 (69.3%) who suggested that 'teachers should recommend references for study from DRI'. 221 (61.1%) were of the view that 'it should be included in the syllabus'.

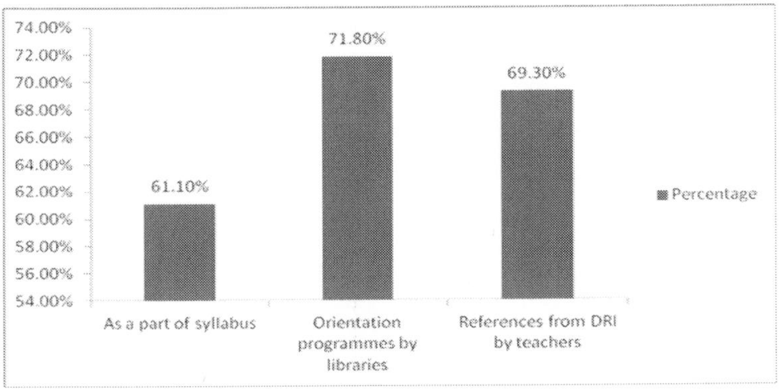

Note: Percentage is out of 362

Fig. 6.6: Suggestions to Improve Efficiency of Using Digital Resources of Information

Table 6.10 shows frequency and percentage of the respondents regarding the suggestions in different age groups. 72.2% respondents in the age group of 20-23 years suggested that 'teachers should recommend references for study from DRI'. On the other hand 78.2% respondents in the age group of 24-27 years and 87.1% of more than 28 years of age suggested that 'libraries should more frequently organize training programmes in this regard'. Thus, it is clear that the students of higher age group expect more from library staff as compared to their younger counterparts who rely more on their teachers to improve their efficiency in using DRI.

Table 6.10: Suggestions to Improve Efficiency of Using Digital Resources of Information by Age

Suggestions to improve efficiency of using digital resources of information	20-23 Years (%age out of180)	24-27 Years (%age out of 151)	>28 Years (%age out of 31)
	R (%age)	R (%age)	R (%age)
It should be included in the syllabus	107 (59.4%)	93 (61.6%)	21 (67.7%)
Libraries should more frequently organize training programmes in this regard	115 (63.9%)	**118 (78.2%)**	**27 (87.1%)**
Teachers should recommend references for study from DRI	**130 (72.2%)**	102 (67.6%)	19 (61.3%)

On analyzing the data for male and female respondents it was found that 78.5% male respondents suggested that 'libraries should more frequently organize training programmes in this regard' (Table 6.11). On the other hand, majority (67.1%) female respondents suggested that 'teachers should recommend references for study from DRI'. The male respondents look forward to library staff in order to improve their efficiency of using DRI whereas female respondents expect the same from their teachers.

Table 6.11: Suggestions to Improve Efficiency of Using Digital Resources of Information by Gender

Suggestions to improve efficiency of using digital resources of information	Male respondents (%age out of 195)	Female respondents (%age out of 167)
	R (% age)	R (% age)
It should be included in the syllabus	120 (61.5%)	101 (60.5%)
Libraries should more frequently organize training programmes in this regard	**153 (78.5%)**	107 (64.1%)
Teachers should recommend references for study from DRI	139 (71.3%)	**112 (67.1%)**

The class-wise frequency distribution and percentage of data with regard to the suggestions are given in Table 6.12. It was clear that 72.3% Master degree students, which constitutes majority of respondents, suggested that 'teachers should recommend references for study from DRI'. However, 87.3% PhD scholars suggested that 'libraries should more frequently organize training programmes in this regard'. It clearly shows that PhD scholars anticipate more on libraries as compared to Master degree students because of lot of research work to be done. On the other hand Master degree students look ahead towards their teachers as the research work in their case is not so intense rather it is comparatively more of classroom teaching, class notes, assignments and exams thus they might be more dependent on their teachers.

Table 6.12: Suggestions to Improve Efficiency of Using Digital Resources of Information by Class

Suggestions to improve efficiency of using digital resources of information	Master degree students (%age out of 260)	PhD scholars (%age out of 102)
	R (%age)	R (%age)
It should be included in the syllabus	159 (61.2%)	62 (60.8%)
Libraries should more frequently organize training programmes in this regard	171 (65.8%)	**89 (87.3%)**
Teachers should recommend references for study from DRI	**188 (72.3%)**	63 (61.8%)

From the Fig. 6.6, Tables 6.10 to 6.12, it is suggested, in support to McCarthy and Vicente that librarians in collaboration with the teachers can work to improve the DIL skills of the students so that they are able to use DRI efficiently. McCarthy opines that in order to make students digital information literate the "role of the teachers needs to be transformed to that of a facilitator or a coach". On the other hand, Vicente the role of new libraries' is to attain efficient skilled use of DRI. In order to achieve this goal, the library should create new partnerships among librarians, faculty and students. He further emphasized that libraries should develop IL courses for the students so that they make optimum use of the whole range of electronic resources (33-44).

6.4 Conclusion

In nut shell, it was found that the majority respondents of all the three age groups and both the sexes exhibited good adoption of DRI in their social lives. However, newspapers are still preferred in print format as compared to online and majority of respondents specifically from both the higher age groups preferred to consult 'employment news' over 'Internet' in order to search a job. In order to improve efficiency of using DRI majority of respondents suggested that 'Libraries should more frequently organize training programmes in this regard'. Thus, it has been realized that DIL is a life-long learning process and acquiring these skills is a major challenge ahead for all of us to ensure an information secure future.

REFERENCES

McCarthy, Lesley. "Digital Literacy- Information Literacy- The New Literacies". *Proceedings ASET Conference: Untangling the Web: Establishing Learning Links. Melbourne, July 7-10, 2002.* Ed. S. McNamara and E. Stacey. N.p., 2002. Web. 18 Oct. 2008.

Vicente, Rocío Serrano. "Academic Libraries and Learning Support in the Electronic Environment." *Estudiossobre Educación* 7 (2004): 33-44. *Dpace.unav.es*. Web. 2 Dec. 2009.

■■■■■

7 DIGITAL INFORMATION LITERACY SKILLS

7.1 Introduction

The digital information literacy skills possessed by the sample in the present study were evaluated by analyzing the data for frequency distribution, percentage, weighted index (W.I.), correlation and association of attributes. Analysis of data for frequency distribution, percentage and W.I. has already been discussed in the previous chapters. The correlation between various DIL skills like digital information competency level score (DICLS), social score, ethical use score and access score of the users was determined by calculating Karl Pearson's correlation coefficient(r). The association of these skills with various variables was calculated by applying Pearson's chi-squared (χ^2) test. The results have been discussed in the following sections.

7.2 Correlation Analysis

The correlation between various DIL skills (as discussed in chapter 2) like DICLS, social score, ethical use score and access score; was calculated to examine the correlation of each score with others. The results of the correlation analysis are given in Table 7.1. DICLS describes competency of an individual in evaluating and using digital information. Social score defines one's adoption of digital resources of information (DRI) in social lives. Access score shows frequency of access and quantum of information accessed from DRI. Ethical use score refers to tendency of ethical use of DRI and due acknowledgement given to the DRI used by the respondents.

Table 7.1: Correlation Analysis

	Social score (r)	Ethical use score (r)	Access score (r)
DICLS	0.335*	0.191*	0.260*
Social Score	-	0.018	0.169*
Ethical Use Score	-	-	0.016

Note:* The correlation is statistically significant at $p \leq 0.05$

7.2.1 Correlation of digital information competency level score with other digital information literacy skills

It is evident from Table 7.1 that DICLS has significant correlation with social score (r = 0.335), ethical use score (r = 0.191) and access score (r = 0.260) and is positively correlated with them. It shows that the respondents having high DICLS access DRI more than others. They are capable of accessing information quickly and in desired proportions than those who possess lower DICLS. They use DRI for non-academic purposes also and ethically use the information from DRI more than others. Such respondents have better acceptance of DIL in their social lives and they use information ethically by giving due credit to the DRI consulted. These capabilities increase with the increase in competency level and the same is true other way round also.

7.2.2 Correlation of social score with other digital information literacy skills

Table 7.1 shows that social score is significantly correlated with access score (r = 0.169). This correlation is positively significant. The respondents with high access score use DIL more for their non-academic requirements than others and vice versa. The respondents capable of accessing information from DRI efficiently have better adoption of DIL in their social lives. As these abilities increase the extent of using DRI in common lives too increases. Social score, on the other hand, has non-significant correlation with ethical use score (r = 0.018). Thus, it is not essential that the respondents who have adopted DIL in their social lives use the DRI ethically too.

7.2.3 Correlation of ethical use score with other digital information literacy skills

The ethical use score as clear from Table 7.1, has statistically non-significant correlation with access score (r = 0.016). It means that it is not necessary that the respondents who efficiently access the DRI tend to use it ethically too.

7.3 Association of Digital Information Literacy Skills with Variables of Study

The association between various DIL skills of respondents like DICLS, social score, ethical use score and access score; and different variables of study was calculated using Pearson's chi-squared test. The results of association of all the variables are discussed in the Table 7.2.

Table 7.2: Association of Digital Information Literacy Skills with Variables of Study

	DICLS		**Social score**		**Ethical use score**		**Access score**	
	χ^2	*df*	χ^2	*df*	χ^2	*df*	χ^2	*df*
Age	7.07*	2	1.10	2	0.83	2	1.77	2
Gender	0.05	2	3.17	2	13.73*	2	2.56	1
University	7.19	4	4.39	2	14.17*	2	8.85*	2
Class	3.73	2	2.89	2	12.21*	1	17.42*	1

Note:* The association is statistically significant at $p \leq 0.05$

Table 7.2 shows that the age of the respondents had statistically significant association with DICLS (χ^2 = 7.07) at p value of less than equal to 0.05 at significant level. Similarly, gender of the respondents was found to be associated with ethical use score (χ^2 = 13.73). The university was associated with ethical use score (χ^2 = 14.17) and access score (χ^2 = 8.85). The class was also found to be associated with ethical use score (χ^2 = 12.21) and access score (χ^2 = 17.42). The association of the variables with the DIL skills is discussed individually in detail in the following sub-sections.

7.3.1 Association of digital information competency level score with variables of study

The association of DICLS with different variables of study namely age, gender, university and class which was calculated using Pearson's chi-squared statistical test is discussed in next page.

Association of digital information competency level score with age

The association between DICLS of users with all the three age groups was studied. Table 7.3 shows that majority of users in all the three age groups have DICLS in the range of 11-13 which shows 'medium' DICL amongst the users of all the three age groups. The respondents in the age group of more than 28 years of age had highest average DICLS of 11.90 which indicates their efficiency in evaluating and using DRI over others. The association was found to be statistically significant as the chi-square (χ^2) value was 7.07 with 2 degree of freedom (df) and significant at p value of less than equal to 0.05. Thus the age of the users was associated with their DICL.

Table 7.3: Association of Digital Information Competency Level Score with Age

Age group	Average score	DICLS (Frequency)		
		Low	*Medium*	*High*
		8- 10	*11-13*	*14-15*
20-23 Years	11.84	34	121	25
Years	11.36	40	104	7
> 28 Years	11.90	5	22	4
Degree of freedom: 2				
Chi-square value: 7.07*				

Note: * The association is statistically significant at $p \leq 0.05$

Association of digital information competency level score with gender

The association between DICLS and gender of users was evaluated using chi-squared statistical test. Table 7.4 shows that majority of male as well as female users have DICLS in the range of 11-13 which shows 'medium' DICL amongst the users of both the sexes. Both male and female respondents had average DICLS of 11.65. However, the association was found to be statistically non significant as the chi-square value was 0.05 with 2 df.

Table 7.4: Association of Digital Information Competency Level Score with Gender

Gender	Average score	DICLS (Frequency)		
		Low	*Medium*	*High*
		8- 10	*11-13*	*14-15*
Male	11.65	42	134	19
Female	11.65	37	113	17
Degree of freedom: 2				
Chi-square value: 0.05				

Note: The association is statistically non significant at $p \leq 0.05$

Association of digital information competency level score with universities

The association between DICLS of the users of all the three universities was observed. Table 7.5 reveals that majority of users in all the three universities have 'medium' DICLS. HAU had maximum average DICLS of 11.75, thus might have comparatively better DICL skills. However, the association between DICL and the universities was found to be statistically non-significant as the chi-square value was 7.19 with 4 df.

Table 7.5: Association of Digital Information Competency Level Score with Universities

Name of the university	Average score	DICLS (Frequency)		
		Low	*Medium*	*High*
		8- 10	*11-13*	*14-15*
HAU	11.75	32	135	17
HPKV	11.25	14	36	3
PAU	11.67	33	76	16
Degree of freedom: 4				
Chi-square value: 7.19				

Note: The association is statistically non significant at $p \leq 0.05$

Association of digital information competency level score with class

The association of DICLS with Master degree students and PhD scholars was found. It is evident from Table 7.6 that majority of Master degree students (170) and PhD scholars (77) have DICLS in the range of 11-13 thus they have 'medium' DICL. PhD scholars have higher average DICLS score of 11.84, therefore might be more efficient than Master degree students in using and evaluating digital information. However, the association between DICL and class was found to be statistically non-significant with chi-square value of 3.73 with df 2.

Table 7.6: Association of Digital Information Competency Level Score with Class

Class	Average score	DICLS (Frequency)		
		Low	*Medium*	*High*
		8- 10	*11-13*	*14-15*
Master degree Students	11.57	63	170	27
PhD Scholars	11.84	16	77	9
Degree of freedom: 2				
Chi-square value: 3.73				

Note: The association is statistically non significant at $p \leq 0.05$

7.3.2 Association of social score with variables of study

The association of social score with different variables of study i.e., age, gender, university and class was calculated using chi-squared statistical test and is discussed below.

Association of social score with age

The association between social score of users with all the three age groups was studied. Table 7.7 shows that majority of users in all the three age groups have social score in the range of 4-6 which shows 'medium' adoption of DIL in their social lives. The respondents in the age group of more than 28 years of age had higher average social score of 6.10; it indicates that they might use DIL for non-academic purposes commonly in their routine life more than other two age groups. The above said association was found to be statistically non-significant with the chi-square value of 1.10 and 2 df.

Table 7.7: Association of Social Score with Age

Age group	Average score	Social score (Frequency)		
		Low	*Medium*	*High*
		1-3	*4-6*	*7-9*
20-23 Years	5.82	13	109	58
Years	5.86	10	88	53
> 28 Years	6.10	0	17	14
Degree of freedom: 2				
Chi-square value: 1.10				

Note: The association is statistically non significant at $p \leq 0.05$

Association of social score with gender

The association between social score and gender of users was evaluated using chi-squared test. Table 7.8 shows that majority of male as well as female users have social score in the range of 4-6 which shows 'medium' level of adoption of DIL in their social lives. Male respondents had higher average social score of 5.92; it indicates that they might use DRI in their common lives more than female respondents. However, the association was found to be statistically non-significant as the chi-square value was 3.17 with 2 df.

Table 7.8: Association of Social Score with Gender

Gender	Average score	Social score (Frequency)		
		Low	*Medium*	*High*
		1-3	*4-6*	*7-9*
Male	5.92	14	107	74
Female	5.78	9	107	51
Degree of freedom: 2				
Chi-square value: 3.17				

Note: The association is statistically non significant at $p \leq 0.05$

Association of social score with universities

The association between social score of the users and all the three universities was observed. Table 7.9 reveals that majority of users in all the three universities have 'medium' social score which shows that they have adopted DIL in their social lives up to a medium level. HAU has maximum average social score of 6.00, which shows better adoption of DIL in social lives by its respondents. However, the association between social score and the universities was found to be statistically non-significant as the chi-square value was 4.39 with df 2.

Table 7.9: Association of Social Score with Universities

Name of the university	Average score	Social score (Frequency)		
		Low	*Medium*	*High*
		1-3	*4-6*	*7-9*
HAU	6.00	11	100	73
HPKV	5.87	5	28	20
PAU	5.65	7	86	32
Degree of freedom: 2				
Chi-square value: 4.39				

Note: The association is statistically non significant at $p \leq 0.05$

Association of social score with class

The association of social score with Master degree students and PhD scholars was calculated using chi-squared statistical test. It is evident from Table 7.10 that majority of Master degree students (157) and PhD scholars (57) have medium social score. PhD scholars have higher average social score of 6.09. This indicates that PhD scholars might use DIL for non academic requirements more than Master degree students. The association between social score and Master degree students along with PhD scholars was found to be statistically non-significant with chi-square value of 2.89 with df 2.

Table 7.10: Association of Social Score with Class

Class	Average score	Social score (Frequency)		
		Low	*Medium*	*High*
		1-3	*4-6*	*7-9*
Master degree Students	5.77	19	157	84
PhD Scholars	6.09	4	57	41
Degree of freedom: 2				
Chi-square value: 2.89				

Note: The association is statistically non significant at $p \leq 0.05$

7.3.3 Association of ethical use score with variables of study

The association of ethical use score with different variables of study i.e., age, gender, university and class was calculated using chi-squared test. The results of the same are discussed below.

Association of ethical use score with age

The association between ethical use score i.e. ethical use of information from DRI by the users with all the three age groups was evaluated. Table 7.11 shows that majority of users in all the three age groups had 'medium' ethical use score which shows ethical use of DRI was at medium level. The respondents in the age group of more than 28 years of age had highest average ethical use score of 7.84. It shows that they might use DRI more ethically giving due acknowledgement to the DRI consulted than other two groups. The association was however found to be statistically non-significant with the chi-square value of 0.83 and df 2.

Table 7.11: Association of Ethical Use Score with Age

Age group	Average score	Ethical use score (Frequency)		
		Low	*Medium*	*High*
		5-6	*7-8*	*9-10*
20-23 Years	7.58	20	152	8
Years	7.44	25	119	7
> 28 Years	7.84	1	29	1
Degree of freedom: 2				
Chi-square value: 0.83				

Note: The association is statistically non significant at $p \leq 0.05$

Association of ethical use score with gender

The association between ethical use score and gender of users was evaluated using chi-squared test. Table 7.12 shows that majority of male as well as female users have 'medium' ethical use score. It shows ethical use of information from DRI was 'medium'. Female respondents had higher average ethical use score of 7.79 confirming more ethical use of digital information by them. The association was found to be statistically significant with chi-square value of 13.73 and 2 df. Thus the gender of the user was statistically significantly associated with the ethical use of information from DRI.

Table 7.12: Association of Ethical Use Score with Gender

Gender	Average score	Ethical use score (Frequency)		
		Low	*Medium*	*High*
		5-6	*7-8*	*9-10*
Male	7.33	36	153	6
Female	7.79	10	147	10
Degree of freedom: 2				
Chi-square value: 13.73*				

Note: * The association is statistically significant at $p \leq 0.05$

Association of ethical use score with universities

The association between ethical use score of the users and universities was evaluated. Table 7.13 shows that majority of users in all the three universities have 'medium' ethical use score which indicates medium level of ethical use of information by them. PAU has maximum average ethical use score of 7.76 which reveals its most ethical use of DRI as compared to the other two universities. The association between ethical use score and the universities was found to be statistically significant, having chi-square value of 14.17 with df 2. It means that ethical use of DRI was associated with the university.

Table 7.13: Association of Ethical Use Score with Universities

Name of the university	Average score	Ethical use score (Frequency)		
		Low	*Medium*	*High*
		5-6	*7-8*	*9-10*
HAU	7.34	35	145	4
HPKV	7.75	5	41	7
PAU	7.76	6	114	5
Degree of freedom: 2				
Chi-square value: 14.17*				

Note: * The association is statistically significant at $p \leq 0.05$

Association of ethical use score with class

The association of ethical use score with Master degree students and PhD scholars was calculated using chi-squared test. It is evident from

Table 7.14 that majority of Master degree students and PhD scholars had medium ethical use score. PhD scholars had higher average ethical use score of 7.91. It shows PhD scholars use information from DRI more ethically giving due credit to the DRI consulted than the Master degree students. The association between ethical use score and Master degree students along with PhD scholars was found to be statistically significant with chi-square value of 12.21 with df 1. It shows that class of study was significantly associated with ethical use of information from DRI by the users.

Table 7.14: Association of Ethical Use Score with Class

Class	Average score	Ethical use score (Frequency)		
		Low	*Medium*	*High*
		5-6	*7-8*	*9-10*
Master degree Students	7.40	43	206	11
PhD Scholars	7.91	3	94	5
Degree of freedom: 1				
Chi-square value: 12.21*				

Note: * The association is statistically significant at $p \leq 0.05$

7.3.4 Association of access score with variables of study

The association of access score with different variables of study i.e., age, gender, university and class was calculated using chi-squared test. The results are discussed below.

Association of access score with age

The association between access score with different age groups was calculated. Table 7.15 shows that majority of users in all the three age groups had 'high' access score which shows their efficiency in accessing DRI was high. The respondents in the age group of more than 28 years of age had highest average access score of 5.81 which indicates that they might be more efficient in accessing digital information quickly and in required quantum as compared to others. However, the association was found to be statistically non-significant with the chi-square value of 1.77 and df 2.

Table 7.15: Association of Access Score with Age

Age group	Average score	Access score (Frequency)		
		Low	*Medium*	*High*
		1-2	*3-5*	*6-7*
20-23 Years	5.67	2	79	99
24-27 Years	5.62	0	64	87
> 28 Years	5.81	1	9	21
Degree of freedom: 2				
Chi-square value: 1.77				

Note: The association is statistically non significant at $p \leq 0.05$

Association of access score with gender

The association between access score and gender of users was evaluated using chi-squared test. Table 7.16 shows that majority of male as well as female users had 'high' access score. It shows their efficiency in accessing information from DRI. Female respondents had higher average access score of 5.80 which indicates that they might access DRI more efficiently than male respondents. The association was found to be statistically non-significant with chi-square value of 2.56 and 1df.

Table 7.16: Association of Access Score with Gender

Gender	Average score	Access score (Frequency)		
		Low	*Medium*	*High*
		1-2	*3-5*	*6-7*
Male	5.55	2	89	104
Female	5.80	1	63	103
Degree of freedom: 1				
Chi-square value: 2.56				

Note: The association is statistically non significant at $p \leq 0.05$

Association of access score with universities

The association between access score of the users and universities was evaluated. Table 7.17 shows that majority of users in HAU and PAU has 'high' access score indicating efficient access of information from DRI by them. HPKV has 'medium' access score. PAU had maximum average access score of 5.91. Thus, confirms their efficiency in accessing information from DRI quickly and in desired amount over the respondents of other two universities. The association between access score and the universities was found to be statistically significant with chi-square value of 8.85 with df 2. It means that access of DRI was associated with the university.

Table 7.17: Association of Access Score with Universities

Name of the university	Average score	Access score (Frequency)		
		Low	*Medium*	*High*
		1-2	*3-5*	*6-7*
HAU	5.60	0	83	101
HPKV	5.30	1	29	23
PAU	5.91	2	40	83
Degree of freedom: 2				
Chi-square value: 8.85*				

Note: * The association is statistically significant at $p \leq 0.05$

Association of access score with class

The association of access score with Master degree students and PhD scholars was calculated using chi-squared statistical test. It is evident from Table 7.18 that majority of Master degree students and PhD scholars had high access score. PhD scholars had higher average access score of 5.90; therefore, they are more capable of accessing DRI. The association between access score and Master degree students along with PhD scholars was found to be statistically significant with chi-square value of 17.42 with df 1. It shows that class of study was significantly associated with access of information from DRI by the users.

Table 7.18: Association of Access Score with Class

Class	Average Score	Access score (Frequency)		
		Low	*Medium*	*High*
		1-2	*3-5*	*6-7*
Master degree Students	5.57	2	127	131
PhD Scholars	5.90	1	25	76
Degree of freedom: 1				
Chi-square value: 17.42*				

Note: * The association is statistically significant at $p \leq 0.05$

7.4 Testing of Hypotheses

Hypothesis 1: There will be no association between digital information competency level and age of users

The association between DICLS and age of users was calculated using Pearson's chi-squared test. The association was found to be statistically significant at χ^2 value of 7.07 with 2 df (Table 7.3). The respondents in the age group of more than 28 years had highest average DICLS of 11.90 which indicates their efficiency in evaluating and using DRI over others. It is clear from Table 4.14 that the respondents of higher age groups are more competent in framing appropriate keywords and this skill increases with age. Table 4.22 shows that the no. of respondents who were able to search the required information from DRI in 'one day' was maximum in the age group of more than 28 years. Thus it is concluded that the age of the user was associated with its DICL. As a result, **the null hypothesis was rejected.**

Hypothesis 2: There will be no association between digital information competency level and gender of users

The association of DICLS and gender of users was found to be statistically non significant with χ^2 value of 0.05 and 2 df. Table 7.4 shows that both male and female respondents had average DICLS of 11.65. Table 4.23 shows that 22.1% male and 22.8% female respondents took 'one day' to search required information from DRI. Thus, it is evident that the difference between the percentage of male and female respondents was negligible which supports the fact that male and female respondents had almost same average DICLS. However, the gender of the user was not associated with their DICL. **Thus the null hypothesis was not rejected**.

Hypothesis 3: There will be no association between digital information competency level and university of users

The association between DICLS and the universities was found to be statistically non-significant as the χ^2 value was 7.19 with 4 df. Table 7.5 reveals that since HAU had maximum average DICLS of 11.75, thus the respondents of HAU might have comparatively better DICL skills than the respondents of HPKV and PAU. The proficiency of the respondents to select an authentic DRI was assessed during the present study and Table 4.31 shows that students from HAU were found to be most competent. Since there was no association between DICL and university of users, therefore, **the null hypothesis was not rejected.**

Hypothesis 4: There will be no association between digital information competency level and class of users

The association of DICLS with class of users was found to be statistically non-significant with χ^2 value of 3.73 with df 2. It is evident from Table 7.6 that PhD scholars have higher average DICLS score of 11.84; therefore they might be more efficient than Master degree students in using and evaluating digital information. The PhD scholars exhibited better understanding of the issue of plagiarism as evident from Table 3.13. Similarly, they were more capable of framing most appropriate keywords (Table 4.16), searching information from DRI with 'ease' (Table 4.21) as compared to the Master degree students. The number of respondents who reported to search information from DRI in 2-5 days was more as compared to Master degree students (Table 4.24). Furthermore, Table 4.28 and 4.32 shows PhD scholars were more competent in evaluating a DRI with regard to authenticity also. However, since the class of the users was not associated with their DICL, **the null hypothesis was not rejected**.

Hypothesis 5: Libraries' programmes are not sufficient to enhance the digital information literacy skills of their users.

The efforts made by libraries like user orientation programmes, special training programmes, assistance provided by library staff and compulsory course on searching information from DRI were found to be insufficient to enhance DIL amongst the students. Tables 5.8 and 5.9 in the section 5.2 shows that library programmes or efforts like various training sessions, user orientation programmes, need based help provided to the users and compulsory course on searching information from DRI made by library staff in this regard were lagging behind as compared to the efforts made by the students at their own level or by their teachers. Thus the programmes conducted by the libraries are not sufficient to enhance the DIL amongst the students. Hence, **null hypothesis was not rejected**.

7.5 Conclusion

It is concluded that the DICL of the respondents was statistically correlated with the social score, ethical use score and access score. However, there was no correlation between social score and ethical use score. On the other hand, the correlation between social score and access score was statistically significant. But ethical use score was not correlated with the access score. The DICL of the users was found to have statistically significant association with the age but was not statistically associated with the gender, university and class. The association of the social score was statistically non-significant with the age, gender, university and class of the respondents. The association of ethical use score was statistically significant with gender, university and class but was statistically non-significant with the age. The association of access score was statistically significant with university and class but it was statistically non-significant with age and gender of subjects. Furthermore, it was also revealed from the present study that the efforts made by the libraries were not sufficient to enhance the DIL skills of the users.

■■■■

8 FINDINGS, CONCLUSIONS, SUGGESTIONS AND RECOMMENDATIONS FOR FURTHER AREAS OF RESEARCH

8.1 Introduction

The present study was conducted to assess the digital information literacy (DIL) skills of Master degree students and PhD scholars of three agricultural universities namely; Chaudhary Charan Singh Haryana Agricultural University, Hisar (HAU); Chaudhary Sarwan Kumar Himachal Pradesh Krishi Vishvavidyalaya, Palampur (HPKV) and Punjab Agricultural University, Ludhiana (PAU). This chapter discusses various findings of the study, conclusions, suggestions and recommendations for further research.

8.2 Findings

The data included demographic information of the respondents like their age, gender, university and class in addition to various questions relevant to evaluation of the DIL skills possessed by them. Findings of the study are presented as under:

1. The total sample of 362 respondents included 184 respondents from HAU, 53 from HPKV and 125 from PAU. It consisted of 260 Master degree students and 102 PhD scholars; 195 male respondents and 167 female respondents; 180 respondents in the age group of 20-23 years, 151 in 24-27 years and 31 of more than 28 years age.

2. Majority of respondents from all the three universities used digital resources of information (DRI) to 'prepare for assignments, exams, etc'. Besides this, 'general awareness' was another major purpose for which DRI was used by an equal number of respondents of HPKV.

3. Majority of respondents opined that DRI are a blessing because 'sufficient amount of information is easily available on any given topic'. However, there were a considerable number of respondents who regarded 'it can be easily copied and used'.

4. Majority of respondents consulted 51% -75% digital references out of the total references, which is not satisfactory. Master degree students consulted more digital references as compared to PhD scholars. Male respondents consulted more digital references than female respondents. The amount of digital references consulted was highest in HAU and lowest in HPKV.

5. E-mail was used most for the purpose of 'communicating information/messages/images amongst friends' by majority of respondents. It was used maximum by the students of more than 28 years of age for 'communicating information/messages/images amongst friends'. The female respondents used e-mail much more than male respondents for the purpose of 'communicating information/messages/images amongst friends'.

6. Majority of the respondents were aware of plagiarism as they would 'use only idea/concept' of a relevant power point presentation available on the internet. The awareness of plagiarism was found to be at maximum in the age group of more than 28 years. The female respondents were much more sensitive regarding the issue of plagiarism as compared to male respondents. PhD scholars of all the three universities were more conscious on the issue of plagiarism as compared to Master degree students.

7. Number of respondents who would 'always' cite the DRI consulted was very less as compared to number of respondents who would cite it 'sometimes'. Majority of respondents i.e. 39% would 'mostly' cite it, which is not satisfactory. The awareness of 'mostly' citing the DRI consulted was found to be maximum in the youngest age group. The female

respondents were more responsive about citing the DRI consulted. PhD scholars of HAU and HPKV were more aware of citing DRI as compared to Master degree students and it was vice versa in PAU.

8. AGRICOLA, AGRIS and CAB CDROM databases were used by majority respondents of all the three universities for 'bibliographic information'; CeRA and Krishiprabha for 'descriptive information' and 'other digital resources available through the Internet' for 'current information'.
9. CeRA was found to be most useful out of all the databases available in the three universities. CAB was second most useful database.
10. The frequency of use of DRI was reported by only 29.6% of respondents as 'daily', which is not adequate. Majority of respondents from PAU used it 'daily'. The 'daily' use of DRI was much more in PhD scholars as compared to Master degree students.
11. 'University library' was found to be most favored place for accessing DRI. The younger age groups viz. 20-23 years and 24-27 years accessed DRI from the 'university library' whereas the older age group i.e. more than 28 years would access it from their 'departments'. Majority of male respondents accessed DRI from the 'university library' whereas female respondents accessed it from their 'home'. The majority of respondents from HAU and HPKV accessed DRI from the 'university library' whereas in PAU it was accessed from 'home'. Majority of respondents who accessed DRI either from their homes or hostels used 'broadband' internet connection.
12. The majority of respondents were able to find 'adequate' information from DRI. This ability was found maximum in the age group of 20-23 years. The female respondents were much more capable of finding 'adequate' information from DRI. The PhD scholars were more capable of finding 'adequate' information.
13. Majority of respondents accessed 'electronic databases subscribed by university library' 'through library website'.
14. In order to search relevant information from the enormous amount available from DRI, the subjects preferred to refine

search by sorting them 'by relevance' instead of going for 'advanced search option' and 'search again using some other keyword'.

15. It was revealed that the subjects preferred 'Google' the most for accessing full text articles as compared to the specialized websites like Google scholar, CeRA and ScienceDirect.com.
16. Around 57% respondents were not able to frame the most appropriate keyword in order to search relevant information from the DRI, which is not satisfactory. Only 43.2% respondents were able to frame most appropriate keyword. This competency was found to increase with the age and was found maximum in the age group of more than 28 years. PhD scholars were much more competent in framing most appropriate keywords as compared to Master degree students. Male respondents were much more capable of framing the most appropriate keyword as compared to female respondents.
17. Majority of respondents were capable of finding 'almost all' and very few were able to find 'all' relevant information from DRI. The ability to find relevant information from DRI was maximum in the age group of 20-23 years. The PhD scholars of HAU were more efficient in finding the relevant information as compared to Master degree students and it was reverse incase of other two universities.
18. Majority of respondents were capable of searching information from DRI 'easily'. The ability of searching information from DRI 'easily' was much more in case of female respondents as compared to male respondents. The respondents from the age group of 24-27 years were able to search information from DRI much more 'easily' than the other two groups. PhD scholars were much more capable of searching information from DRI 'easily' as compared to Master degree students.
19. It was observed that majority of respondents took five days or even less than that to search information from DRI, which is satisfactory. There were maximum respondents in the age group of 24-27 years who required '2-5' days to search information. There were more female respondents as well as PhD scholars who were capable of searching information from DRI in five days or even less than that.

20. There were adequate respondents who were able to evaluate the authenticity of a website. This ability was found high across all age groups, universities, both the sexes and classes.
21. The websites of educational institutions were considered by the majority, as most authentic for research based information.
22. The inability to frame appropriate keywords was regarded as the major reason of finding irrelevant information from DRI. This was followed by helplessness to use 'advanced search' option.
23. 'Virus infected computers' was reported as one of the major infrastructural or hardware related problems followed by 'slow internet connectivity' faced by the users while accessing DRI.
24. Library staff was regarded 'very helpful' by around one-third of respondents whereas there were sufficient respondents who considered them 'helpful'. Males regarded them more 'helpful' as compared to female respondents. The library staff was regarded as 'very helpful' by maximum respondents from HPKV and the least by HAU.
25. The library staff was perceived as 'average trained' by majority respondents. Library staff was regarded as 'well trained' by majority of respondents in the age group of 20-23 years whereas maximum respondents in the age group of 24-27 years and of more than 28 years of age considered them as 'average trained'. Respondents from HAU and HPKV considered their library staff as 'well trained' whereas the majority from PAU held them 'average trained'.
26. 'Self learning from internet/manuals, etc' was regarded as most helpful factor in using DRI as against 'orientation from library staff'.
27. 'References of digital resources of information given by teachers' were considered as most important factor in referring information from DRI as compared to various efforts made by the library staff. However, more than 58% of the respondents reported that less than 50% of digital references were given by their teachers. The percentage of digital references given by the teachers was found to be least in PAU and highest in case of HPKV.

28. Majority of respondents would 'check the status from website and reserved the ticket online' incase of railway reservation. Similarly, majority respondents used ATM for withdrawal of money from bank and would 'search for relevant information from the specific website' for the purpose of going abroad. This shows that majority respondents of all the three age groups and both the sexes exhibited good adoption of DRI in their social lives.
29. Newspapers are still preferred in print format by majority of respondents as compared to online. It was also revealed that majority of respondents preferred to consult 'employment news' over 'Internet' in order to search a job. The majority of respondents in the lower age group i.e. 20-23 years consulted Internet to search a job as compared to users of higher age groups i.e. 24-27 years and more than 28 years age, who preferred to consult 'employment news'. The respondents from PAU consulted 'Internet' whereas it was 'employment news' in HAU and HPKV. Majority of Master degree students and PhD scholars as well as male and female respondents preferred to consult 'employment news'.
30. In order to improve efficiency of using DRI, majority of respondents suggested that 'libraries should more frequently organize training programmes in this regard'. The youngest age group, female respondents and Master degree students suggested that 'teachers should recommend references for study from digital resources' to improve their efficiency for using DRI. Whereas both the older age groups, male respondents and PhD scholars suggested that 'libraries should more frequently organize training programmes in this regard'.
31. The digital information competency level (DICL) of the respondents was found to be statistically correlated with the social score, ethical use score and access score of the respondents.
32. There was no correlation between social score and ethical use score of the respondents.
33. The correlation between social score and access score was statistically significant.
34. The ethical use score was not correlated with the access score.

35. The DICL of the users was found to have statistically significant association with the age.
36. The DICL of the users was not statistically associated with the gender, university and class.
37. The association of the social score was statistically non-significant with the age, gender, university and class of the respondents.
38. The ethical use score was statistically not associated with the age of the user.
39. The association of ethical use score was statistically significant with gender, university and class.
40. The access score of the respondents was statistically not associated with age and gender of subjects.
41. The association of access score was statistically significant with university and class of the respondents.
42. Efforts made by the libraries were not sufficient to enhance the DIL skills of the users.

8.3 Conclusions

The present study is concluded as below:

1. Majority of the students used DRI to the amount of 51%-75% of the total resources consulted. The extent of use of DRI was not found satisfactory and needs to be popularized amongst them.
2. The users were capable of accessing adequate and relevant information 'easily' in a span of five days or even less than that from DRI. They also possessed the ability of evaluating a DRI regarding its authenticity. But they were found less competent in framing appropriate keywords and using advanced search option which leads to irrelevant information.
3. In order to accomplish the training needs of the users with regard to the use of DRI, the library needs to reframe its existing training programmes. Firstly, because various such training sessions, user orientation programmes and compulsory course on searching information from DRI; were found to be insufficient. Secondly, the users were dependent more on self learning and references given by teachers as compared

to the need based assistance from the library staff. However, the number of references given by teachers from DRI was not satisfactory. More exhaustive efforts may be made by the teachers in this respect.

4. The understanding of the users with regard to ethical issues in using DRI was found to be in nascent stage as although they exhibited sufficient awareness about plagiarism but still considerable number of respondents were found to be irresponsive towards acknowledging the DRI used and regarded DRI a blessing as 'it can be easily copied and used'. Much more needs to be done to educate them on this aspect.
5. The users have successfully adopted DIL skills in their social lives and the same was found to be better implemented by the users of younger age group.
6. The majority of respondents of all the three universities considered their library staff as 'helpful' in using DRI whereas it should be 'very helpful' in order to enable the efficient and optimum use of DRI by the users and improvement of their DIL skills.
7. The library staff of HAU and HPKV was considered as 'well trained' whereas it was regarded as 'average trained' in PAU.
8. Statistically significant correlation was found between DICL and social score, ethical use score as well as access score of the respondents. Social score and access score were also significantly correlated.
9. DICL skills of the users were statistically associated with age of the users. The respondents of more than 28 years of age had highest average DICLS thus were more efficient in evaluating and using DRI over others.
10. Ethical use score was statistically associated with gender, university and class. Ethical use of information, giving due credit to the DRI consulted, was confirmed more in case of female respondents and PhD scholars over male respondents and Master degree students respectively. PAU had maximum average score which reveals its more ethical use of DRI as compared to the other two universities.
11. Access score was found to be statistically associated with university and class. PAU had maximum average access score thus confirms their efficiency in accessing information from

DRI quickly and in desired amount over the respondents of other two universities. Similarly, PhD scholars too were found to be more capable of accessing DRI than Master degree students.

8.4 Suggestions

The following points are suggested consequent to evaluating the DIL skills of Master degree students and PhD scholars of HAU, HPKV and PAU in the present study:

1. The use of DRI needs to be more popularized amongst the users by educating them regarding its benefits. This may be done on priority by creating awareness through user orientation/training programmes, institutional website, references given by teachers and librarians and as an important part of the course curriculum.
2. The skills of framing appropriate keywords and using advanced search option needs to be developed amongst the users comprehensively. For this purpose, basic as well as advanced level training courses may be organized regularly. Furthermore, the evaluation of these skills prior and subsequent to such training courses should be considered a mandatory feature.
3. The Library and Information Science (LIS) professionals need to play pivotal role in inculcating DIL skills amongst the users with constant efforts for their improvement. Since it was revealed from the present study that the efforts made by libraries viz various training sessions, user orientation programmes, need based assistance from the library staff and compulsory course on searching information from DRI; were insufficient to enhance the DIL skills of the users; so it is extremely essential to thoroughly review all such efforts. In this regard the libraries may reorganize the structure of their training/user education programmes by constantly updating the contents in order to keep pace efficiently and effectively with requirements of the users. Secondly, the training programmes should also focus on improving those DIL skills of the users in which they are already lacking. Thirdly, the contents of the course should also be regularly revised with prime focus on developing DIL skills amongst users.

4. During the course of this study, the LIS professionals in all these three universities started teaching library science and its services with emphasis on the use of DRI subscribed by these libraries as a part of course on technical writing for Master degree students and PhD scholars which was earlier on confined to teaching of compilation of bibliography only. Though accessing DRI is included in the curriculum of this course but the students are not introduced to the concept of DIL specifically. It is crucial to educate the students about the theory of DIL and the importance of these skills to use DRI effectively so that the students turn out to be digitally information literate citizens coping up with ever exploding word of digital information.

5. It is worth mentioning here that the library and its services part of the technical writing course for Master degree students and PhD scholars was being taught as a non-credit course in PAU. During the course of this study, it was upgraded as a credit course and it has become a part of the overall grading of the students now. Consequently, the students consider it more seriously and will be able to improve their DIL skills. It is already a credit course in HPKV but it is still a non-credit course in case of HAU. It is recommended that this course may also be made a credit course at HAU to improve their DIL skills.

6. The users need to be educated regarding the ethical use of DRI to improve their understanding and importance about the responsive use of digital information by giving due acknowledgement to the DRI consulted. This issue may be included as a mandatory part of each and every training programme imparted by the libraries. Secondly, it may also be included as an essential component of the course curriculum. Last but not least, teachers may create awareness regarding this along with other teaching assignments.

7. The users of higher age groups also need to be encouraged to adapt the use of DIL in their social lives.

8. The library staff needs to change their attitude and extend maximum help to the users in making optimum use of DRI. Resultantly, it would improve the DIL skills of the users.

9. The library staff of PAU needs to constantly update their own skills so that they are able to impart proper training to the users.
10. It was evident that the students rely more on teachers to improve their DIL skills, therefore, a mandatory collaborative approach among teachers and librarians may be established. The classroom teaching and the references given by teachers from DRI along with various efforts made by libraries might enhance the DIL skills of the users.

8.5 Recommendations for Further Areas of Research

Under-graduate students were purposely not involved in the present study. However it is recommended that a study may be conducted on evaluating DIL skills of UG students also so that the required improvements may be made. Similarly, a study on assessment of DIL skills possessed by the teachers and LIS professionals is also recommended as it would consequently sharpen their DIL skills which in turn will lead to improvement of the DIL skills of students. The department wise DIL skills analysis will draw better results of students from various departments.

This study is limited to agricultural universities of Punjab, Haryana and Himachal Pradesh only. It is recommended that a study may be conducted on all the agricultural universities of India to obtain comprehensive results. Not only in agricultural universities but steps may be taken to evaluate DIL skills of the students, teachers and LIS professionals of all the schools, colleges and universities in phased manner. As a next step comparisons with other universities at global level can be made which would certainly bring out scope of improvement of DIL skills amongst the students, teachers as well as LIS professionals of the country, thus achieving a complete digital information literate India.

■■■■

BIBLIOGRAPHY

American Library Association. Progress Report on Information Literacy: An Update on the American Library Association Presidential Committee on Information Literacy: Final Report. March 1998. Chicago: American Lib. Assn., 1998. Web. 15 Nov. 2009.

Andretta, Susie. "From Prescribed Reading to the Excitement or Burden of Choice: Information literacy: Foundation of E-learning." *Aslib Proceedings: New Information Perspectives* 57.2 (2005): 181-90. *Emerald.* Web. 17 Sept. 2008.

Appleton, Leo. "LolliPop for Learning Resources: Information Literacy Staff Training within Further Education." *Journal of Librarianship and Information Science* 42.3 (2010): 191-98. *Sage Journals Online*. Web. 29 May 2012.

Association of College and Research Libraries. *Information Literacy Competency Standards for Higher Education.* Chicago: American Lib. Assn., 2000. Web. 18 Oct. 2008.

Avdic, Anders and Anders Eklund. "Searching Reference Databases: What Students Experience and What Teachers Believe that Students Experience." *Journal of Librarianship and Information Science* 42.4 (2010): 224-35. *Sage Journals Online*. Web. 28 May 2012.

Bawden, David. "Information and Digital Literacies: A Review of Concept." *Journal of Documentation* 57.2 (2001): 218-59. Print.

Best, John W. *Research in Education.* New Jersey: Prentice-Hall, 1959. Print.

Bewick, Laura, and Sheila Corrall. "Developing Librarians as Teachers: A study of their Pedagogical Knowledge." *Journal of Librarianship and Information Science* 42.2 (2010): 97-110. *Sage.* Web. 24 Sept. 2011.

Bridgland, Angela, and Martha Whitehead. "Information Literacy in the 'E' Environment: An Approach for Sustainability." *Journal of Academic Librarianship* 31.1 (2004): 54–59. *Science Direct*. Web. 9 Oct. 2011.

Brown, Cecelia, and Lee R. Krumholz. "Integrating Information Literacy into the Science Curriculum."*College and Research Libraries* 63.3 (2002): 111-23. Print.

Bruce, Christine Susane. *The Seven Faces of Information Literacy.* Adelaide: Auslib Press, 1997.Web. 29 May 2010.

Busha, Charles H, and Stephen P. Harter. *Research Methods in Librarianship: Techniques and Interpretation.* New York: Academic Press, 1980. Print.

Chakraborty, Susmita, Anup Kumar Das, and B. K. Sen. International Conference on Information Management in Digital Libraries. Aug. 2-4, 2006: Use of Indian Digital Libraries by Biomedical Practitioners in West Bengal Region: A Study. Kharagpur: n.p., 2006. Openmed.nic.in. Web. 18 Oct. 2008.

Chauhan, S. K., Prem Chand, and T. A. V. Murthy. "Information Literacy for Indian Academicians: INFLIBNET Initiatives." *Journal of Library and Information Science* 31.1 (2006): 45-52. *Sage Journals Online*. Web. 28 May 2012.

Covello, Stephen. *A Review of Digital Literacy Assessment Instruments*. N.p.: Syracuse Univ, 2010. Web. 4 April 2011.

Crawford, John. "The Use of Electronic Information Services and Information Literacy: A Glasgow Caledonian University Study." *Journal of Librarianship and Information Sciences* 38.1 (2006): 33-44. Print.

Dadzie, Perpetua S. "Information Literacy: Assessing the Readiness of Ghanaian Universities." *Information Development* 23.4 (2007): 266-77. *Sage Journals Online*. Web. 23 July 2009.

Doherty, John J., Mary Anne Hansen, and Kathryn K. Kaya. "Teaching Information Skills in the Information Age: The Need for Critical Thinking." *Library Philosophy and Practice* 1.2 (1999): 1–12. *E-prints in Lib. and Information Sciences.* Web. 29 July 2009.

Ercegovac, Zorana. "What Students Say They Know, Feel, and Do About Cyber-Plagiarism and Academic Dishonesty? A Case Study." *Proceedings of the American Society for Information Science and Technology* 42.1 (2005). *Wiley Interscience*. Web. 18 Nov. 2009.

Fahmy, Engy I., and Nermine M. Rifaat. "Middle East Information Literacy Awareness and Indigenous Arabic Content Challenges." *International Information and Library Review* 42 (2010): 111-23. *Science Direct*. Web. 17 June 2011.

Farkas, Meredith. "Participatory Technologies, Pedagogy 2.0 and Information Literacy." *Library Hi Tech* 30.1 (2012): 82-94. *Emerald*. Web. 10 Feb. 2013.

Flaspohler, Molly R. "Information Literacy Program Assessment: One Small College Takes Big Plunge." *Reference Services Review* 31.2 (2003): 129-40. *Emerald*. Web. 14 Oct. 2009.

Fowler, Clara S., and Elizabeth A Dupuis. "What Have We Done? TILT's Impact on Our Instruction Program." *Reference Services Review* 28.4 (2000): 343-48. *Emerald.* Web. 12 Oct. 2009.

Gainer, Jesse. "Critical Thinking: Foundation for Digital Literacies and Democracy." *Journal of Adolescent and Adult Literacy* 56.1 (2012): 14–17. *Wiley.* Web. 11 Feb. 2013.

Gantz, John F., et al. The Diverse and Exploding Digital Universe: An Updated Forecast of Worldwide Information Growth by 2011 (IDC White Paper). N. p., 2008. Web. 18 Oct. 2008.

Gedam, P. B., and A. T. Agashe. International Conference on Academic Libraries. Oct. 5-8, 2009: Information Literacy Competencies and Programmes in India: Proceedings. Delhi: N.p., 2009. Print.

Gupta, Santosh. *Research Methodology and Statistical Techniques.* New Delhi: Deep and Deep Pub, 2008. Print.

Hadengue, Veronique. "E-learning for Information Literacy: A Case Study." *Library Review* 54.1 (2005): 36-46. *Emerald.* Web. 4 Sept. 2009.

Hepworth, Mark. "Approaches to Providing Information Literacy Training in Higher Education: Challenges for Librarians." *New Review of Academic Librarianship* 6.1 (2000): 21–34. *Taylor & Francis Online.* Web. 24 Aug. 2009.

Johnston, B., and S. Webber. "Information Literacy in Higher Education: A Review and Case Study." *Studies in Higher Education* 28.3 (2003): 335–52. *ingentaconnect.* Web. 20 Sept. 2009.

Joint, Nicholas. "Information Literacy Evaluation: Moving Towards Virtual Learning Environments." *The Electronic Library* 21.4 (2003): 322-34. *Emerald.* Web. 20 Aug. 2009.

Joint, Nicholas. "Teaching Intellectual Property Rights as Part of the Information Literacy Syllabus." *Library Review* 55.6 (2006): 330-36. *Emerald.* Web. 9 Sept. 2009.

Kennedy, Cathleen. Measuring Information Literacy: The 'Tool Literacy' Variable. 1998. Google. Web. 29 July 2009.

Kokic, Ivana Batarelo. "Information Literacy for Future Teachers." *World Journal of Education* 2.1 (2012): 45-55. *Sciedu Press.* Web. 4 Feb. 2013.

Kothari, C. R. *Research Methodology: Methods and Techniques.* 2nd ed. New Delhi: Wishwa Prakashan, 1990. Print.

Krishnamurthy, M. "Digital Libraries Gateway for Library and Information Sciences: A Study." *SRELS Journal of Information Management* 39.3 (2002): 245-54. Print.

Lampert, Lynn D. "Integrating Discipline Based Anti-plagiarism Instruction Into the Information Literacy Curriculum." *Reference Services Review* 32.4 (2004): 347-55. *Emerald.* Web. 31 Oct. 2009.

Lanham, Richard A. "Digital Literacy." *Scientific American* 272.3 (1995): 160-61. Print.

Leigh, Jennifer S. A., and Cynthia A. Gibbon. "Information Literacy and the Introductory Management Classroom." *Journal of Management Education* 32.4 (2008): 509-30. *Sage Journals Online.* Web. 22 Oct. 2009.

Lindstrom, J., and D.D. Shonrock. "Faculty–Librarian Collaboration to Achieve Integration of Information Literacy." *Reference and User Services Quarterly* 46.1 (2006): 18–23. Print.

Lorenzo, George. Catalysts for Change: Information Fluency, Web 2.0, Library 2.0, and the New Education Culture. N.p., 2007. Web. 18 Dec., 2008.

Maharana, Bulu, and Champeswar Mishra. "A Survey of Digital Information Literacy of Faculty at Sambalpur University." *Library Philosophy and Practice* 9.2 (2007). *E-prints in Lib. and Information Sciences.* Web. 18 Oct. 2008.

Martin, Allan. "Essential e-literacy." *Connected* 9 (2003): 22–4. Web. 28 Oct. 2009.

Martin, Allan. "Literacies for the Digital Age." *Digital Literacies for Learning.* Eds. Allan Martin and Dan Madigan. London: Facet, 2006. 3–25. Web. 4 June. 2009.

Martin, Allan, and Jan Grudziecki. "DigEuLit: Concepts and Tools for Digital Literacy Development." *ITALICS* 5.4 (2006): 249-67. *Directory of Open Access Journals.* Web. 4 April 2011.

McCarthy, Lesley. "Digital Literacy- Information Literacy- The New Literacies". *Proceedings ASET Conference: Untangling the Web: Establishing Learning Links. Melbourne, July 7-10, 2002.*Ed. S. McNamara and E. Stacey. N.p., 2002. Web. 18 Oct. 2008.

McShane, Ian. "Public Libraries, Digital Literacy and Participatory Culture." *Discourse: Studies in the Cultural Politics of Education* 32. 3 (2011): 383-97. *Routledge.* Web. 14 July 2012.

Media Awareness Network. Digital Literacy in Canada: From Inclusion to Transformation. A Submission to the Digital Economy Strategy Consultation. 2010. Web. 4 April 2011.

Negi, M. S. Research and Statistical Techniques in Library and Information Science. New Delhi: Shree Pub and Distributors, 2005. Print.

Nelson, Klara, Marcy Courier, and Gilbert W. Joseph "Teaching Tip: An Investigation of Digital Literacy Needs of Students." *Journal of Information Systems Education* 22.2 (2011): 95-109. *EBSCOhost.* Web. 20 Nov. 2012.

Owusu-Ansah, Edward K. "Information Literacy and the Academic Library: A Critical Look at a Concept and the Controversies Surrounding it." *Journal of Academic Librarianship* 29.4 (2003): 219-30. *Science Direct.* Web. 3 June 2011.

Prasher, R. G., and R. K. Sharma. ICT Based Information Management in Indian Libraries: Festschrift in Honour of Prof. (Dr.) K C Sahoo. New Delhi: Bookwell, 2012. Print.

Ray, Kathryn, and Joan Day. "Student Attitudes towards Electronic Information Resources." *Information Research* 4.2 (1998). *Google.* Web. 29 July 2009.

Rockman, H. "Successful Strategies for Integrating Information Literacy into the Curriculum." *Integrating Information Literacy into the Higher Education Curriculum: Practical Methods for Transformation.* Eds. Ilene

F. Rockman et al. San Francisco, CA: Jossey-Bass, 2004. 47-91. Web. 18 Oct. 2010.

Romani, Cobo, and Cristobal Juan. Strategies to Promote the Development of E-competencies in the Next Generation of Professionals: European and International trends. Google. Web. 6 Feb 2013.

Rowlands, Ian, and David Bawden. "Digital Libraries: A Conceptual Framework." *Libri* 49.4 (1999): 192-202. Print.

Savolainen, R. "Network Competence and Information Seeking on the Internet: From Definitions towards a Social Cognitive Model." *Journal of Documentation* 58.2 (2002): 211–26. Print.

Seamans, Nancy H. "Student Perceptions of Information Literacy: Insights for Librarians." *Reference Services Review* 30.2 (2002): 112-23. *Emerald.* Web. 29 Oct. 2009.

Sharma, K. R. *Research Methodology.* New Delhi: National Pub House, 2002. Print.

Sieberhagen, Elsabe Anee. The Design and Development of a Digital Information Literacy Programme for an Academic Library. Diss. U of So. Afr., 2005.N.p., 2005. U of So. Afr. Web. 18 Oct. 2008.

Society of College, National and University Libraries. Information Skills in Higher Education: Briefing Paper, Prepared by the SCONUL Advisory Committee on Information Literacy. London: SCONUL, 1999. Web. 17 Mar. 2009.

Spitzer, Kathleen L., Michael B. Eisenberg, and Carrie A. Lowe. *Information Literacy: Essential Skills for the Information Age*. Santa Barbara, CA: Libraries Unlimited, 1998. Print.

Sridhar, M. S. Information Behaviour of Scientists and Engineers: A Case Study of Indian Space Technologists. New Delhi: Concept Publishing Co., 1995. Print.

Stubbings, Ruth, and Alan Brine. "Reviewing Electronic Information Literacy Training Packages." *ITALICS* 2.1 (2003). *Directory of Open Access Journals.* Web. 20 Jan. 2013.

Taylor, John K, and Cheryl Cihon. *Statistical Techniques for Data Analysis.* 2nd ed. Florida: Chapman and Hall/CRC, 2004. Print.

Todd, Ross. "Information Literacy in Electronic Environments: Fantasies, Facts, and Futures." *Proceedings of the International Association of Scientific and Technological University Libraries Conferences* (2000). *Google.* Web. 18 Nov. 2009.

Tryon, Patricia J. Slagter van. "The Instructional Designer's Role in Forming University- Community Partnerships in Digital Literacy." *Tech Trends* 57.1 (2013): 52-8. *Springer Link.* Web. 16 Feb. 2013.

Vicente, Rocío Serrano. "Academic Libraries and Learning Support in the Electronic Environment."*Estudiossobre Educación* 7 (2004): 33-44. *Dpace.unav.es*. Web. 2 Dec. 2009.

Welsh, Teresa S., and Melissa S Wright. *Information Literacy in Digital Age: An Evidence –Based Approach.* New Delhi: Chandos Publishing, 2010. Print.

Williams, Peter. "Opinion: Against information literacy." *Library and Information Update* 5.7–8 (2006): 20- 22. Web. 28 Aug. 2009.

Wilson, Myoung Chung. "To Dissect a Frog or Design an Elephant: Teaching Digital Information Literacy Through the Library Gateway". *63rd IFLA General Conference, Amsterdam, Aug. 31- Sep. 5, 1997. INSPEL* 32.3 (1998): 189-95. *Intl. Federation of Lib. Assn. and Inst.* Web. 18 Sept. 2008.

Zurkowski, Paul G. *The Information Service Environment: Relationships and Priorities.* Washington, DC: National Commission on Libraries and Information Science, 1974. *Google.* Web. 10 Jan. 2009.

Annexure

QUESTIONNAIRE

This questionnaire is a part of a research study Digital Information Literacy among Master Degree Students and PhD Scholars of Agricultural Universities: A Case Study of Punjab, Haryana and Himachal Pradesh. Your reply to this questionnaire is highly valuable and extremely important to assess and evaluate digital information literacy amongst the students of agricultural universities. Information provided by you will be kept confidential and used for academic purpose only.

DEMOGRAPHIC INFORMATION

1. Name (Optional):__________________________________

Please tick mark (√) the appropriate option (question numbers 2 -3)

2. Age Group:

   a) 20 - 23 years __________ b) 24– 27 years_________

   c) 28 years and above ______________

3. Gender: a) Male ________ b) Female __________

4. Name of the University: ____________________________

5. Department: ____________________________________

6 Class:____

USING DIGITAL RESOURCES OF INFORMATION

7. For what purpose do you use digital resources of information? *Please check all the answers which apply.*

Sr. No.	Purpose	Your Answer (√)
1	Preparing for assignments, exams, etc.	
2	Writing research paper (for journal, conference, etc.)	
3	For general awareness	
4	Thesis related work	
5	Preparing class notes	
6	Doing some new experiments	
7	Any other purpose, please specify__________	

Please tick mark (√) the appropriate answer(s) (question numbers 8-9)

8. On an average, how many references out of the total references consulted by you are from digital resources of information?

   a) 1%-20% _________ b) 21%-40% ________________

   c) 41%-60% ________ d) 61%-80% ________________

   e) 81-100%__________

9. How many references given by your teachers are of digital resources of information?

   a) 1%-20%_________ b) 21%-40%______________

   c) 41%-60% _________ d) 61%-80% ______________

   e) 81-100%_________

10. In your opinion the abundantly available information via digital resources of information is a blessing because *(Please check all the answers which apply)*

Sr. No.	Digital resource is a blessing because	Your Answer(√)
1	It can be easily copied and used	
2	Sufficient amount of information is easily available on any given topic	
3	A person can keep update on all topics of interest	
4	Any other please specify_________________	

11. You use e-mail for the following purposes *(Please check all the answers which apply)*

Sr. No.	Purposes for using e mails	Your Answer(√)
1	Receiving alerts of latest articles from various databases of full text journals	
2	For discussion forums	
3	Communicating information/messages/images amongst friends	
4	Chatting	
5	Any other, please specify________________	

12. You have found that PowerPoint presentation on the topic of your credit seminar is available on Internet. What will you do? *Please check all the answers which apply.*

    a) Download and use it __________________

    b) Make changes in language and design and then use it________

    c) Use only idea/concept_________________

    d) Any other, please specify__________________________

13. Do you cite the website or digital resource of information used by you? *Please tick mark (√) the appropriate option.*

    a) Always____________ b) Almost every time___________

    c) Mostly_____________ d) Sometimes_________________

    e) Never_____________

14. Rate the usefulness of the following digital resources of information? *Please tick mark (√) the appropriate option given against each.*

Sr. No.	Digital resources of information	Frequency of use				
		Very usefu l	Useful	Satisfactory	Of little use	Of no use
1	AGRICOLA					
2	AGRIS					
3	CAB					
4	CeRA (Consortium for e-Resources in Agriculture)					

5	Krishiprabha					
6	Other digital resources of information available through Internet					
7	Any other, please specify____________					

15. For what type of information do you use the following digital resources of information? *Please tick mark (√) the appropriate option given against each.*

Sr. No.	Digital resources of information	Descriptive information	Statistical information	Current information	Conceptual information	Bibliographic information	Any other, please specify
1	AGRICOLA						
2	AGRIS						
3	CAB						
4	CeRA (Consortium for e-Resources in Agriculture)						
5	Krishiprabha						
6	Other digital resources of information available through Internet						
7	Any other, please specify __________						

ACCESSING DIGITAL RESOURCES OF INFORMATION

16. How frequently do you access digital resources of information? *Please tick mark (√) the appropriate option.*

a) Daily ___________ b) Once in two days ______

c) Weekly _________ d) Fortnightly______

e) Monthly_________ f) Any other, please specify________

17. How do you access digital resources of information? *Please check all the answers which apply.*

Sr. No.	Resource	Your Answer(√)
1	Electronic databases subscribed by your university library	
2	Electronic databases other than subscribed by your university library	
3	Other digital resources of information available through Internet	
4	Any other, please specify____________	

18. How do you access the digital resources of information subscribed by your library? *Please check all the answers which apply.*

    a) Directly through URL of respective digital resource of information ________________________________

    b) Through your library website______________________________

    c) Any other, please specify __________________________________

19. From where do you access digital resources of information? *Please check all the answers which apply.*

    a) University Library____________________________________

    b) Department ____

    c) Cybercafe ____

    d) Home (From your own laptop/personal computer) _________

    e) Hostel ____

    f) Any other, please specify ______________________________

If you access digital resources of information from your home or hostel, what type of Internet connection do you have? *Please tick mark (√) the appropriate option(s).*

a) Broadband____ b) Leased line_____

c) Wi-fi _____ d) Dial up______

e) Cable _____________

f) ISDN (Integrated Services Digital Network) ________

g) DSL (Digital Subscribers Line) ______

h) GPRS (General Packet Radio Service) ___

i) Any other, please specify______________

20. Please specify the type of Internet connection available in your university. *Please check all the answers which apply.*

    a) Broadband__________ b) Leased line__________

    c) Wi-fi _____ d) Dial up_____________

    e) Cable ____________

    f) ISDN (Integrated Services Digital Network) ________

    g) DSL (Digital Subscribers Line) ______

    h) GPRS (General Packet Radio Service) ___

    i) Any other, please specify________________

21. The amount of information accessed by you from digital resources of information is *(Please tick mark (√) the appropriate option)*

    a) Adequate _________________

    b) Inadequate __________________

    c) Very inadequate __________________

22. You find an article which is very important for your research work but that article is available on payment. What would you do to access the article? *Please tick mark (√) the appropriate option.*

    a) Pay the price of the article____________________

    b) Try to find a person/library who has access to that article _____________________

    c) Try to contact the author and request for a free copy of the article_________

    d) Try to find the abstract of the article _____________________

    e) Ignore it and search for some other freely available article ______________________________

    f) Any other, please specify ____________________________

SEARCHING INFORMATION FROM DIGITAL RESOURCES OF INFORMATION

23. You are given an assignment to find information from any database mentioned in the question number 15 on a topic of your interest. Suppose you receive enormous number of references on searching, how will you proceed further? *Rank your answers as 1 for highest rank and 10 for lowest rank.*

Sr. No.	Options	Your Rank
1	Consult all the references one by one	
2	Consult the first 10 or 20 references	
3	Randomly consult few of them	
4	Refine your search by sorting chronologically	
5	Refine your search by sorting them 'by relevance'	
6	Refine your search by sorting highly cited	
7	Going for advanced search	
8	Search again using some other keyword	
9	Give up the search	
10	Any other, please specify________________	

24. Which out of the following resources will you use for searching full text article on any topic of your interest? *Rank your answers as1 for the highest rank and 6 as the lowest.*

Sr.No.	Options used	Your Rank
1	CeRA	
2	ScienceDirect.com	
3	Google	
4	Google Scholar	
5	Altavista	
6	Any other please specify ________________	

Please tick mark (√) the appropriate option(s) (question numbers 25-28)

25. In order to search information on 'Nutritional value of apple' from any of the CDROM databases subscribed by your university library, which is the most appropriate keyword out of the following?

    a) Apple-nutrition________________________________

    b) Apple and nutrition____________________________

    c) Nutrition in apple_____________________________

    d) Apple or nutrition_____________________________

    e) Apple not nutrition____________________________

    f) Any other, please specify ________________________

26. How much relevant information do you find while searching required information from digital resources of information?

    a) All ___ b) Almost all ___

    c) Little ___ d) Very little ___

    e) None ___

27. Searching information from digital resources of information is:

    a) Very easy ____ b) Easy ____

    c) Somewhat easy ____ d) Not easy____

    e) Difficult____

28. On an average how many days do you usually take to search information on a given topic?

    a) One day ___________ b) 2-5 days __________

    c) One week _______ d) 8-10 days __________

    e) More than 10 days____

29. Rank the importance of the following options with regard to searching information from digital resources of information. *Rank your answers as 1 for the highest rank and 8 as the lowest.*

Sr. No.	Statements	Your Rank
1	Assistance from library staff whenever need arises	
2	Information provided in 'User Orientation Programmes'	
3	Compulsory course on searching information from digital resources of information	
4	Special training sessions organized by the library	
5	References of digital resources of information given by your teachers	
6	Help and instructions for using digital resources of information by teachers	
7	Own efforts	
8	Any other please specify_________________	

SELECTION

Please tick mark (√) the appropriate option(s) (question numbers 30-31)

30. Which of the following do you regard as most authentic digital resource of information for 'state-wise sex ratio in India'?

    a) www.india.gov.in __________

    b) www.cyberjournalist.org.in ______

    c) www.infochangeindia.org ____

    d) www.censusindia.gov.in _________

31. Which out of the following options is best resource to find out 'total number of agricultural universities and research institutes in India'?

    a) (i) www.icar.org.in _____________

       (ii) www.indiaedu.com _____________

       (iii) www.cyberjournalist.org.in ____

       (iv) www.career.webindia123.com ____

    b) Why? Give reasons. Rank your answers as 1 for the highest rank and 7 for the lowest.

Sr. No.	Reasons	Your Rank
1	Accessible	
2	Authentic	
3	Highly used	
4	Regularly updated	
5	Relevant	
6	User friendly	
7	Any other, please specify________________	

32. Rank the following type of websites for authenticity in order to access research based information. *Rank your answers as 1 for the highest rank and 5 as the lowest.*

Sr. No.	Options used	Your Rank
1	Government websites (*.gov.in)	
2	Educational institutions websites (*.edu)	

3	Commercial websites (*.com)	
4	Voluntary research/other such organization websites (*.org)	
5	Any other please specify ________________	

PROBLEMS FACED

33. Rank the reasons given below for getting irrelevant information. *Rank your answers as 1 for the highest rank and 8 for the lowest.*

Sr. No.	Reasons	Your Rank
1	Unable to frame appropriate key words	
2	Sufficient number of digital resources of information are not available	
3	Journals covered in the databases are not useful	
4	Unable to use 'advanced search' option	
5	Unable to refine the search	
6	Library staff is not helpful in searching information from digital resources	
7	Proper training is not imparted in this regard	
8	Any other, please specify____________________	

34. What problems do you face while accessing the digital resources of information? *Rank your answers as 1 for the highest rank and 7 as the lowest.*

Sr. No.	Options used	Your Rank
1	Slow Internet connectivity	
2	Unavailability of printer	
3	Inadequate number of computers	
4	Saving data on pen drives or CDs	
5	Virus infected computers	
6	Inadequate power back-up	
7	Any other please specify ________________	

ROLE OF LIBRARY/UNIVERSITY

Please tick mark (√) the appropriate option(s) (question numbers 35-36)

35. How much helpful is the library staff in using digital resources of information?

    a) Very helpful ____________ b) Helpful ___________

    c) Not helpful _________________

36. Does your syllabus include something on how to search information from digital resources of information?

    a) Yes __________________ b) No _______________

37. Which of the following options usually help you in using digital resources of information? *Rank your answers as1 for the highest rank and 6 as the lowest*

Sr. No.	Options used	Your Rank
1	Orientation from library staff	
2	Orientation from teachers	
3	Orientation from database companies/suppliers/aggregators, etc.	
4	Self learning from Internet/manuals, etc.	
5	Asking the peer group/seniors/colleagues, etc.	
6	Any other please specify ________________	

38. How much well trained is your library staff in helping you to use digital resources of information? *Please tick mark (√) the appropriate option.*

    a) Well trained ___________ b) Average trained ____________

    c) Not-trained ______________

DIGITAL INFORMATION LITERACY IN SOCIAL LIVES

Please tick mark (√) the appropriate option(s) (question numbers 39-41)

39. You get a chance to attend an international conference for example at National Academy of Agricultural Research Management, Hyderabad. How will you proceed?

    a) Immediately go to railway reservation counter and reserve a seat ____________

b) You would request your friend to immediately go to railway reservation counter and reserve a seat for you ____________

c) Check the reservation status from railway website first and then go to railway reservation counter to reserve a seat ________________

d) Check the reservation status from railway website first and then reserve a seat online ___

e) Any other, please specify__________________________

40. You come to know that today is the last day to deposit semester fee. What will you do?

a) Withdraw money from bank using withdrawal form ________

b) Withdraw money from bank using cheque _______________

c) Withdraw money using ATM card ___________________

d) Request the bank staff for immediate help in withdrawal of money ______

e) Any other, please specify_____________________________

41. Which format do you prefer to read the newspapers?

a) Print____________ b) Online__________

c) Both____________

42. You wish to go abroad for further studies. How will you proceed? *Please check all the answers which apply.*

Sr. No.	Options used	Your Answer (√)
1	Go to immigration consultancy	
2	Consult your friends	
3	Consult your teachers	
4	Search for relevant information from the specific websites	
5	Any other, please specify____________	

43. In order to search suitable job for yourself, which of the following do you consult. *Please check all the answers which apply.*

Sr. No.	**Options**	**Your Answer (√)**
1	Employment news	
2	Newspapers	
3	College notice board	
4	Internet browsing	
5	Visiting the websites of the concerned companies	
6	Enquiring the placement cell of your institution	
7	Register online in various job sites	
8	Placement agency	
9	Any other option, please specify ____________________	

SUGGESTIONS

44. In order to improve efficiency of the students in using information from digital resources of information, what out of the following do you suggest. *Please check all the answers which apply.*

Sr. No.	**Suggestions**	**Your Answer(√)**
1	It should be included in the syllabus	
2	Libraries should more frequently organise training programmes in this regard	
3	Teachers should recommend references for study from digital resources of information	
4	Any other please specify________________	

■■■■